W9-AGV-189

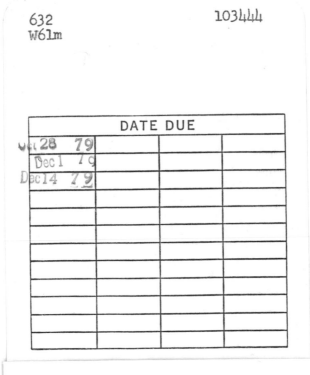

Microbial Plant Pathology

Studies in the Biological Sciences
General Editor: *Professor V. Moses*

The Biology of Insects: *C. P. Friedlander*
Freshwater Biology: *L. G. Willoughby*

Microbial Plant Pathology

P. J. Whitney

Pica Press   New York

Published in the United States of America in 1977 by
PICA PRESS
Distributed by Universe Books
381 Park Avenue South, New York, N.Y. 10016

Library of Congress Catalog Card Number: 76-20406
ISBN 0-87663-722-5

Printed in Great Britain

Contents

Chapter 1
Introduction

IT IS likely that parasitism has existed since almost the origin of life: the living cell contains so many easily available nutrients it is virtually certain that at a very early stage of evolution organisms became specialized as parasites. We have no fossils of these very early organisms, but fossil fungi have been found in pre-cambrian cherts nearly 2000 million years old. By the time plants started to colonize the land, about 430 million years ago, there is good fossil evidence that their parasites adapted to this new environment with them.

1.1 Plant disease in human history

Whilst man was nomadic, plant disease would probably have affected him very little, but when he started to cultivate plants as crops (about 9000 years ago) he provided ecological changes greatly favouring the disease organisms.

The growth of large numbers of the same type of plant in dense stands provides ideal conditions for the rapid spread of disease. The repeated growing of crops in the same place often causes the build up of large populations of pathogens. Also the breeding of crops to give higher yield has frequently produced crops more susceptible to disease. Finally, the increasing speed and ease of movement of people and plant material about the world has resulted in the rapid spread of disease over very large distances.

The Bible gives us an early record of plant disease altering the course of human history. In about 1800 BC Joseph interpreted the dream of one of the Egyptian Pharaohs in which seven fat kine were eaten up by seven lean kine and seven good ears of grain were eaten up by seven 'blasted' ears. (The Hebrew word *yeraqon* is better translated as 'yellowing' or 'rust' rather than 'blast'.) Joseph said it meant seven good years would be followed by seven years of famine when the east wind would destroy the grain. The east wind is interesting, because the east wind in Spring tends to bring rain and the climate of the eastern Mediterranean is thought to have been wetter then than it is now. The rain brought by the east wind would provide just the conditions for a rust epidemic. Grain provided during the good years was stored for use during the forecast famine years. The Israelites too had a rust epidemic; many of them moved to Egypt, where their skills as craftsmen could earn them food and a good standing in society. Over the years their position in society declined until most of them became slaves. This was the origin of the situation from which the Israelites were liberated when Moses lead them out of Egypt.

In more recent history there have been major alterations brought about by plant disease. When Peter the Great, Czar of Russia, attacked the Turks in an attempt to win ice-free ports for Russia, he met little opposition. Constantinople, and the Turkish empire was in what has been called its 'Tulip Period'. Growing tulips was a craze which left little time for affairs of state and the control of armies. Czar Peter marshalled his troops at Astrakhan and swept down into the Volga delta. This was a rich rye growing area and his troops bought and plundered rye flour for the men and rye hay for the horses. In August 1722, the first horse died of the 'blind staggers' and within hours the first men were dying of 'holy fire'. This was the result of poisoning by alkaloids present in the ergots of the fungus *Claviceps purpurea*, a pathogen of rye. Czar Peter's troops were rapidly killed as they had no idea of the cause of their malady. In all, some 20 000 troops died that Autumn, and caused his attack to fail. Had it succeeded it is likely that the whole of Europe would have fallen to his enlarged empire.

Over the years, ergot poisoning has killed many people despite our knowledge of the cause of this disease. On 12 August 1951, in France, a case of poisoning occurred and others followed. The cause was not immediately recognized as ergotism and in the end the greed of the three men who had sold flour knowing it to be contaminated with ergots resulted in four deaths, thirty two cases of insanity, and over two hundred cases of poisoning from which people recovered.

The potato famine in Ireland in the middle 1840s, was due to a number of causes. Much of Ireland was owned by absentee landlords who wanted a good return from the land they owned. The land was rented out to agents who sub-divided the land and in turn let it, often to other agents who further sub-divided the land into plots small enough to be rented by peasant farmers. This meant practically all the grain grown in the country went to pay the landlord and his agents. The small amount of land left had to be used for a very high yielding crop to provide food enough for the peasants. The very high yielding crop of potatoes therefore became the staple diet. In good years there was an excess which could not be stored for very long because of fungal attack and so was discarded. Fungi, including *Phytophthora infestans* grew on these discarded potatoes and infected the following year's crop. 1844 looked as if it was going to be another excellent year, with very rapid early growth in the warm damp spring. The weather then turned cold and wet and the plants started to wilt, yellow and die. By the Winter of 1845, many people were dying of starvation, but those surviving had great hopes for the following year. 1846 was if anything worse. People died by the thousand of typhoid, typhus and dysentery, which they resisted whilst well fed, but starvation left them prey to these diseases engendered by their living conditions. One million people died and two million people emigrated, causing the population of Ireland to fall by over one third.

Forty five years later it was found that spraying potato plants with copper salts was an effective way of controlling the pathogen. This discovery did not however mean that there would never be another potato famine caused by *Phytophthora*.

In 1915 the German potato crop was enormous, three times the amount needed. To avoid frost damage the potatoes were stored in warehouses. However most warehouses were already full of war materials, so the basements of schools and public buildings were filled with potatoes. The price of potatoes was so low that often it was not economic to transport them and they were discarded. Soon the warm basements containing potatoes began to smell as the potatoes began to rot. Students and teachers were given 'stench vacations' whilst the potatoes were removed and dumped. The Spring of 1916 was good and growth of potato plants was rapid but soon spores from the *Phytophthora* growing on the discarded and dumped potatoes infected the new season's growth and they began to die. Agriculturalists knew that if the potato plants were sprayed with copper compounds the crop could be saved, but all the copper in Germany was going for shell casings and field telephone wire. Most of the 1916 potato crop was lost and all the grain grown was commandeered for the army. Several hundred thousand German civilians died of starvation during the winter of 1916-17, and the news of friends and relatives starving reduced the morale of the German troops and a major forward push intended for 1917 never took place. This loss of morale undoubtedly contributed to their defeat in 1918.

Plant diseases have not only affected population movements, they have also affected social habits. In 1870 coffee was a popular British beverage, and very little tea was drunk. Ceylon exported over one hundred million pounds of coffee in that year. At about this time coffee rust was first observed. The rust epidemic was so disastrous that by 1886 production was down to eighteen million pounds per year, by 1889 to five million pounds and by 1892 there was scarcely a coffee plant in Ceylon. About a quarter of the planters were completely bankrupt and left the country. The remainder, most in severe financial difficulties, were induced by altered government tariffs to start the slow change-over to tea production. Between 2500 and 4000 tea bushes are planted per acre. Hence over a thousand million tea seedlings had to be transported and planted, followed by a five year wait until full production was reached. This enormously difficult change over was made and Britain became an island of tea drinkers, not coffee drinkers, as the result of a coffee rust epidemic.

1.2 The development of the science of plant pathology

One of the first people we can consider as a botanist about whom we have written records was Theophrastus of Eresus. He was a Greek who studied together with Aristotle under Plato. He lived from about 370-286BC. When Aristotle died he gave all his library and botanic garden to his younger colleague Theophrastus. This was the beginning of a school which had of the order of 2000 students.

The writings of Theophrastus are a mixture of careful personal observation and erroneous beliefs widely held at that time. His work on the anatomy of plants was of a far higher order than anything that was done until the Renaissance, one thousand eight hundred years later. He made accurate observations on the response of different plants to disease and the effect of climate and character of land on the likelihood of infection. He recognized many different diseases, but was totally unaware that they were caused by micro-organisms. He still believed in spontaneous generation. It was generally accepted that fungi and bacteria associated with disease arose spontaneously from the dying plant and this belief also persisted until the Renaissance.

The first step towards a germ theory of disease was the statement by Porta in 1588 that the black spores of mushroom were 'seeds', but he had no proof of this. In 1665 Robert Hooke saw the teleospores of *Phragmidium* and believed they were seed pods. However, he still thought spontaneous generation occurred but the fungus produced these structures for further propagation.

Malpighi also investigated fungal spores and believed that fungi grew only from them or from fragments of themselves, but he too lacked proof. Joseph Patton in 1705 made the bold statement that fungi reproduced by eggs or seeds and caused the dangerous mouldiness disease of plants that occurred in humid greenhouses during winter. He thought the humidity hatched the eggs or seeds in minute crevices in the plant surface and recommended greenhouses be kept drier in winter to avoid this. His ideas seem to be one of the first scientific attempts at disease prevention.

In 1729, Pier Antonio Micheli published his accounts of work largely on higher plants but including important work on fungi. He conducted an ingenious series of experiments with species of agarics. He scattered their spores on dead leaves which he incubated at selected sites in the woods, and observed their growth. He also cultured *Mucor, Botrytis and Aspergillus* on freshly cut surfaces of melons, quinces and pears. He varied the conditions of his experiments, made replicates and repetitions and used non-inoculated controls in a series of highly scientific experiments. These experiments gave clear cut and convincing results that these spores (or seeds as he called them) consistently produced crops of their own kind. He obviously understood that spores float in the air, as he put down his occasional aberrant result to contamination by airborne spores. This was one of the first steps to overthrow the theory of spontaneous generation.

Smut is the cause of considerable cereal loss, both by reduced yield and by contaminated grain being unpalatable. The losses from the two smut diseases (loose smut and stinking smut) were so serious that in 1750 the Academy of Arts and Sciences in Bordeaux offered a prize for the best investigation into smuttiness of wheat. M. Tillet noticed that there were two sorts of smut, one in which the ears were not filled with grain, but with brownish balls of an evil smelling black powder − we now call this stinking smut or bunt. The other produced an ear in which only the central rachis and the outer glumes were present and the rest was just a mass of black powdery spores. In both cases the vegetative parts of the wheat plant seem quite healthy. M. Tillet carried out some experiments using an experimental design of field trial that would be very acceptable even today. He used 120 plots with a number of replicates of each treatment, including controls. He repeated his experiments for three years varying the time of sowing and the cultural methods. He showed that the application of the black dust from smutty wheat to smut-free seeds greatly increased the incidence of the disease in the resulting crop, and also that washing the smutty grain before sowing reduced the incidence of the disease. He of course won the prize.

However, he had not realized that the black dust was spores of a parasitic fungus. He thought it contained a poisonous principle that could partly be antidoted or washed away, and that the black dust was a product of poisoned wheat plants.

At about this time, another disease was being investigated. Again in France, a Dr Thiuller realized that 'Holy Fire' was in fact poisoning due to eating the ergot of rye, but he had no idea how or why the rye plants produced these ergots. It was not until 1853 that Louis Tulasne discovered that ergots were the fruiting bodies of an organism quite distinct from the rye plant.

Disease control was already being practised in the mid-seventeenth century even though the cause of infection was not known. In Rouen in France it was noticed that the black stem rust of wheat was particularly severe where there were barberry bushes growing in the hedges. This view gained ground until in 1660 a law was passed passed ordering farmers to destroy every barberry bush. Eradication of the barberry certainly reduced the incidence of the disease, even though nobody knew why. We now know that the barberry is the alternative host for the fungus *Puccinia graminis* and that it is on this host that it undergoes the sexual part of its life cycle. By 1726 the first barberry eradication law was passed in America and the fight against barberry and black stem rust has continued ever since. It has never been totally successful and black stem rust is still a major problem in some areas.

The development of plant pathology was at this time still being hampered by the theory of spontaneous generation. In 1705 de Tournefort classified plant diseases into those caused by internal factors and those caused by external factors. Amongst the external factors he included mouldiness. Hales writing of hops in 1727 stated that he considered the seed of mould from mouldy hops might cause the infection in successive years. Felice Fontana working in 1766 and Giovanni Targioni-Tazzetti in 1767 came independently to the conclusion that cereal grain rusts were caused by microscopic parasitic plants. However they lacked any experimental proof of these advanced ideas.

Men such as these were in a minority and there was still considerable support for the theory of spontaneous generation and the idea that fungi were the product of disease, rather than the other way about. This is not so unreasonable when one remembers that in the majority of plant diseases the production of visible reproductive structures by the pathogen occurs fairly late in the disease cycle, some time after other disease symptoms are apparent.

It was not until the nineteenth century that the theory of spontaneous generation was really ousted. In 1807 Benedict Prevost published a treatise on 'The immediate cause of loose smut and other diseases of plants – and methods of disease control'. This was a brilliant piece of work far ahead of its time, but it received many adverse comments. Prevost investigated the smut disease over a period of ten years. He gave accurate and detailed descriptions of the disease at all stages, and demonstrated that the 'globules' were 'gemmae' or spores of a cryptogam. He showed that the fungus was the immediate cause of the disease, but that it can only incite disease under suitable conditions. He was not however successful in his attempts to study the development of the pathogen from its infection of the seed to its sporulation in the host fruit, but did show that treatment of the seed with copper salts prior to sowing greatly reduced the incidence of the disease. Few if any of his conclusions have had to be modified in the light of modern findings but nearly half a century had to elapse before his work was readily accepted.

From 1842 onwards, the Tulasne brothers carried out detailed studies on a number of fungi and showed clearly the morphology and natural relationship of both saprophytic and parasitic fungi to their substrate or host. They knew of the work of Prevost and thought highly of it.

Heinrich Anton de Bary also worked on plant pathogens and although he only knew of Prevost by brief review articles his findings were substantially the same. In 1855 de Bary founded a school, his students working along lines laid down by him. This school was a very active influence until his death in 1888, and the many fungi isolated and characterized by him and his students still bear his name as a specific name or suffix. It was de Bary who finally demonstrated beyond doubt that *Phytophthora infestans* was the causal agent of potato blight.

Scholars were considerably behind agriculturalists in disease control. In 1755, M. Tillet partially controlled bunt of wheat with lime and in 1824 Robertson controlled peach mildew (which he considered to be a fungus) by repeated applications of sulphur and soap in water. In 1845 Edward Tucker satisfied himself that the mildew on leaves of vines in his glass house was similar to the mildew of peaches and applied sulphur and slaked lime. The causal fungus was for a time named *Oidium tuckeri* after him. So that when de Bary first published in 1853 it was only the most conservative of the scientific population he had to convince.

From this time onwards progress in the field of plant pathology proceeded apace. In the 1860s the school of de Bary contributed much useful work on members of the Peronosporaceae and their work on *Sclerotinia sclerotiorum* started the physiological approach to plant pathology. 1858, saw the first book on plant pathology that had been written with an understanding of the role of fungi in disease. This was by Kühn who had succeeded where Prevost had failed in observing the growth of bunt through the wheat root up to the developing ear.

Brefeld in the late 1870s pioneered the technique of growing fungi in pure culture. These techniques, modified by Koch, Petri and others, is basically the technique used today. At about this time the first investigations were carried out on the diseases of trees. The treatise by Hartig in 1882 was for many years the most comprehensive work in this field.

Professor Millardet of the University of Bordeaux was particularly interested in *Phylloxera* infestations and the downy mildews of vines, pioneering methods which have since become some of our main approaches to the control of plant disease. He selected plants genetically resistant to *Phylloxera* and used fungicidal sprays to control the downy mildew. The fungicidal mixture containing copper sulphate and slaked lime has since become known as Bordeaux Mixture. The use of this mixture which is effective against a wide range of pathogens has had as great an effect on plant pathology as the use of Penicillin has had on human pathology.

The rediscovery of Mendel's laws of genetics in 1900 greatly stimulated the breeding of plants for disease resistance. Biffen (1905) reported that the disease resistance of wheat to *Puccinia glumarum* was inherited as a Mendelian recessive character. This work and the work that followed was aimed largly at combining the disease resistant character of some of the wild varieties of crop plants with the high yield of their cultivated relatives. This work has contributed greatly to our control of many, but by no means all diseases.

It was much more difficult to prove the rôle of bacteria in disease than that of fungi, and it was some twenty five years after the conclusive demonstration of diseases caused by fungi that the same thing was shown for bacteria. Pasteur started things moving by demonstrating the bacterial fermentation of broth. He and Koch paved the way still further by showing in 1876 that anthrax is due to a bacterial infection. Burrill in Illinois studied the blight of pear and apple from 1878 onwards. He demonstrated that a bacterial organism was constantly present and that he could incite the disease by direct inoculation. Other workers soon showed a wide range of diseases were due to bacterial infection.

Virus diseases have of course proved even harder to explain. Mayer, in 1886 demonstrated the experimental transmission of tobacco mozaic virus (TMV) by the injection of juice from a diseased plant into a healthy plant. Infectivity of this juice was not lost by heating to 60°C but was slowly lost at 80°C. He sought to discover the causal organism but not surprisingly he was unable to culture it. He still however concluded that the disease was due to a bacterium. At about the same time Smith was studying peach yellows virus in the USA. He was only able to transmit this disease by grafting, but despite this concluded that it was similar to tobacco mozaic virus. In 1892, Dimitri Ivanowski working in St Petersburg passed juice from a diseased tobacco plant through a bacteria-proof filter and found the juice still induced mozaic symptoms, but he did not realize the significance of what he had done. He still thought of the causal agent as either a bacterium or a toxin. Six years later a Dutch microbiologist, M.W. Beijernick, studied tobacco mozaic by the best bacteriological methods available and again found the causal agent would pass through bacteria-proof porcelain filters and also that it would pass through agar gel. Further, he demonstrated that the infective agent increased in the host and was therefore unlikely to be a toxin. He concluded the disease could not be due to a microbe and referred to it as a *'contagium vivum fluidum'* and sometimes as a 'virus'.

1.3 The situation today

It has been estimated that in the early 1950s America lost three billion (3×10^9) dollars worth of agricultural produce per year due to plant disease. This was 7 per cent of the entire national product, and it is likely that the loss is similar if not higher today. It is certain that losses in the tropics and in less developed countries are higher due to climate and poorer agricultural conditions.

Control of plant disease is a continual battle and although we have an increasing armoury of weapons the enemy still attacks us, often where we least expect it. A continual campaign has to be waged against the diseases, some of which have already been mentioned, but there are a vast number more, many of very great economic importance. For example, South American leaf blight (SALB) of rubber caused by *Dothidella ulei* is controlled only by an expensive system of multiple grafting to produce resistant trees. Bacterial blight of cotton caused by *Xanthamonas malvacearum* is controlled by growing resistant varieties and by seed dressing. There are many other examples of infections of non-food crops but it is the diseases of food crops which are perhaps the most damaging both in terms of economics and of human suffering.

Many of these diseases are spreading. The white stripe virus of rice was known only in Japan in the 1950s. Since then it has increased its geographic range to Cuba and Venezuela, and recently it has been found in Panama and Florida. In Cuba and Venezuela it has reduced the total rice yield by between 20 per cent and 50 per cent. *Puccinia polysora*, a tropical rust, has also spread rapidly since World War 1. It is now especially serious in West Africa where it can cause losses of the order of 40 per cent.

Plant diseases are not always harmful to man. If the disease results in little loss of yield then it is not considered severe from a human standpoint, and if it does not affect crop plants it has little or no economic impact. In some instances disease of weed plants may even be agriculturally advantageous. For example the wilt of Persimmon trees in Southwest USA may be considered serious by some people in that their loss reduces food and cover for wild life and increases the risk of soil erosion. Ranchers on the other hand view the death of the trees with pleasure as the trees reduce the effective area of grazing for cattle. In other cases disease is deliberately used to the advantage of man. The inoculation of pine trees with a variety of pathogenic fungi to increase the flow of resin is just one instance. There are a few examples of plant diseases which in the past have been harmful to mankind which are now used for his benefit. *Claviceps purpurea* is now cultivated so that the alkaloids can be extracted from the ergots and used for medicinal purposes.

How can the incidence of plant disease be reduced? There are a number of ways: (i) cultural methods, (ii) breeding resistant plants, (iii) chemical treatment. It is clear that all these methods cost money and effort, and even if perfect control were obtained the cost would still have to be considered as a loss due to disease.

This loss can be reduced by not applying a treatment when it is not needed and by applying the right treatment when it is needed. What does this mean exactly? Obviously there is no need to apply a treatment to crops that are healthy and are going to remain healthy. Less obvious is the fact that the agriculturalist does not want to spend more on crop protection than is returned in improved yield. So plants that only become lightly infected with little loss of yield should not have expensive or time-consuming treatments applied to them. It is, however, better to verge on the side of caution and apply treatments that may perhaps be unnecessary rather than risk serious damage or the loss of a crop. As our knowledge increases we hope not only to have better control measures, but to know more precisely when and how to use them.

1.4 The economics of disease

If we are to make studies of diseased plants and determine what is the loss due to disease we must know what is the healthy norm. This should mean that every plant reaches its full stature and produces the maximum yield of which it is capable. In reality this is not possible, and often is not even desirable. The agriculturalist is concerned with obtaining the maximum profit from his land. It is therefore normal practice to sow plants closer together than will allow them to reach their maximum stature. Each plant will then yield less than its full potential but the loss is more than made up by the increased number of plants, so that the yield per unit area of land is higher. Further, if some of the plants die then the reduced competition

between plants will allow the survivors to grow more freely and to have a larger
yield, partly compensating for the loss of some of the crop. Application of fertilizer
is also very rarely optimal because farmers and horticulturalists wish only to apply
fertilizer where the increased yield more than pays for the cost of fertilizer and its
application. High levels of fertilizer usually increase the yield only slightly more
than moderate applications.

The United States Department of Agriculture defines normal as the yield that
occurs in good years over extended areas. This is reasonable, but there will still be
even in good years a certain percentage loss due to disease. They then go on to state
that a perfect crop is one that exceeds this by 10 per cent. This is obviously not
reasonable as it assumes the loss in good years due to disease is less than 10 per cent,
something not proven, and it implies that it is impossible to have a yield more than
10 per cent greater than their norm i.e. more than perfect. Not infrequently crops
do yield more than 10 per cent above such a norm. In Germany the norm is taken
as the average yield in those years in which there is no known adverse condition,
climatic or pathological.

Having defined 'normal' by whatever methods, how much do we have to spend
in crop protection to achieve this and is it economic? In a study made by Schultz,
in 1938, he found from statistics that there were inelastic demand curves for ten
major crops. That is, demand did not increase in proportion to the fall in cost and
decrease in proportion to the rise in cost. For example:

Corn (Maize)   0·5 per cent decrease in production = 1 per cent increase in price

Wheat   0·1 per cent increase in production = 2 per cent decrease in price

This means over-production depresses the price more than the increase in yield so
that total income for the farmer falls and under-production tends to make the
farmers' income rise. It would appear at first sight that it is more profitable to under-
produce, so why spend money on crop protection to increase yield? The answer to
this is twofold. Firstly, an individual farmer increasing or decreasing his yield will
have a negligible effect on the total crop produced. He therefore gains the benefit of
increased yield without the disadvantage of decreased price. Also crop protection is
at its most useful in the cases of high incidence of disease when prices will be high
and any yield is better than none. Secondly, the world as a whole is under-
producing so that as transport improves the effect of localized over-production
(local, that is, on a global scale) will be reduced. This is aside from the moral issue
of growing more food to provide starving people with enough to eat.

Crops are subject to two forms of risk from plant diseases: the risk of epidemic
disease completely destroying the crop, and the risk of low levels of infection which
do not become epidemic, i.e. acute and chronic disease. The acute form may be the
more obvious and the most damaging in the short term, but chronic infection
probably causes more total loss in yield. The chronic disease can lie anywhere
between two extremes, one in which the disease causes relatively little damage to
each individual plant and reduces its yield only slightly, but affects a large proportion of the entire crop, or secondly, a disease which causes severe damage or even

the death of an individual plant but affects only a few individuals in the crop. (A severe infection affecting a large proportion of the crop would be an acute infection or epidemic.)

As well as wanting to know the level of disease in a crop at one instance in time, the present, we want to know what is likely to happen in the future. Such a prediction can be made in two ways. The first method uses direct observation of the diseased plant and the relationship between growth of the host and parasite. This sort of prediction is pretty subjective and depends on the skill and experience of the plant pathologist and his assessment of the cultural conditions, future climatic conditions, etc. This method is useful in giving an immediate prediction and is often vital if treatment is to be applied in time to save the crop. Secondly, there are observations over a period of time together with the objective scoring of symptoms and numbers of infected plants. This is less subjective but it is slow. It is however useful where a disease is spreading from a limited number of foci in a large area of crop e.g. forests. It is also useful in scientific experiments where saving the plants on an experimental plot is not the desired aim.

1.5 What is disease?

This might seem a strange question to ask. Until now a fairly general idea of disease has been adequate, but now we require something more specific. We must be clear at the outset that it is not just the presence of a micro-organism in a plant. Non-photosynthetic micro-organisms vary in their mode of life: some are saprophytic living on dead organic material, others are parasitic and grow on living organic material, causing damage. Still others inhabit living tissue, gaining benefit from it and giving benefit in return; these are called symbiotic. There are of course micro-organisms which lie on the borderlines between these groups, saprophytes that can under suitable conditions become parasitic, and parasites which live part of their life cycle saprophytically.

Furthermore, the amount of pathogen in an infected plant does not necessarily indicate how severely the plant is diseased. Some varieties of a particular crop are much more tolerant of the pathogen and are therefore less diseased by it than are other varieties. Nor are the visible symptoms always a reliable guide to the degree of disease as there are many metabolic processes occurring in plants the disturbance of which results in no visible symptoms. Symptoms do however, provide some guide as to the degree of infection and are useful when comparing plants of the same variety suffering from the same disease.

Disease can perhaps best be defined as an alteration to the metabolism of an organism resulting in debility of the individual concerned. The disability can be caused in a variety of ways; (i) mineral deficiencies, (ii) presence of toxic compounds, (iii) damaging cultural practice, (iv) attack by animals, (v) attack by micro-organisms. It is this last group which particularly concerns us in this book, but many of the principles laid down are equally applicable to diseases caused in other ways.

Chapter 2
Spread and transmission of disease

2.1 The production of inoculum

Anything produced so that a pathogen can initiate a new infection is inoculum. There is therefore a wide range of different types of inoculum.

Viruses have such a reduced structure that one tends to think only of the complete particles when considering a virus. This however is the inoculum, the dispersal phase. In the metabolically active phase the various components of the virus are separated so that the nucleic acid can direct the synthesis of new viral subunits in the host cell. With the exception of the spore- and cyst-forming types, there is no difference in the structure of the dispersive and vegetative phases of the life cycle of most bacteria. In contrast to this, fungi produce many different kinds of inoculum. Frequently a single fungus may produce two, three, four or even more distinct forms of inoculum during its life cycle.

Vegetative mycelium can act as inoculum, a fact frequently used in the laboratory sub-culturing of fungi but in its normal environment the mycelium is often too delicate to act as efficient inoculum. It is too easily destroyed by such adverse conditions as drying, ultra-violet light, heat and various chemicals. Many fungi therefore produce specialized resting mycelia with thick walls to resist drying, pigmentation to reduce sensitivity to ultra-violet light, and large food reserves. Vegetative and resting mycelia frequently form an important source of inoculum in vegetatively propagated plant material and in the soil, often associated with plant debris. Developments of resting mycelia are the specialized sclerotia produced by some fungi. These are composed of tightly packed masses of cells with similar characters to resting mycelia but with even thicker walls and more pigmentation and often greater food reserves. Sclerotia are produced by fungi such as *Sclerotinia sclerotiorum* and *Claviceps purpurea* and micro-sclerotia by *Verticillium albo-atrum* and *V. dahliae.*

Some fungi produce structures called rhizomorphs, which are large bundles of hyphae. These grow out from centres of infection through a non-infectable area until a new host is reached. Rhizomorphs can be up to several metres in length and are produced by fungi such as the honey cap or shoe string fungus (*Armillaria mellea*) which attacks and kills a wide range of trees, especially conifers, and *Fomes lignosus* a root rot of rubber trees. Rhizomorphs are highly effective means of establishing a new infection as they are able to draw nutrients from the old host whilst they are establishing themselves in the new host but they are limited in the distance they can spread an infection.

Spores of one sort or another are a very common form of fungal inoculum. It is not surprising when one considers the ecological situation that fungi have evolved these tiny propagules. Unless there is a specialized mechanism for transporting the spores to a suitable host, as in the case of vector-spread diseases, the inoculum is spread at random and the chances of a spore landing on a suitable host are often very small. (They are of course much greater in a monoculture such as an agricultural crop.) This means that a very large number of spores must be produced to give any chance of the fungus establishing a new infection. As the source of nourishment is often limited the size of the spore must be small, and the small size makes dispersal much easier.

It is interesting to note that the parasitic flowering plants have responded to similar ecological pressures in a similar manner to that of the fungi. The inoculum (seeds) of the flowering plant parasites are produced in extremely large numbers and are very small. Although they are not produced in such large numbers as fungal spores, nor are they as small, the trend is clearly the same.

In order to improve dispersal many fungi produce specialized fruiting structures to place the spores where they can easily be carried away by air currents, water splash or a vector. So that the amount of vegetative tissue is not excessive a single support structure usually bears many spores. The complexity of the structure varies from a single hypha bearing a chain of spores at its tip to the highly complex structures produced by the polypores and gill-bearing fungi. Many fungi have a number of different spore-producing bodies at different stages in their life cycles. Often asexual spores are formed on relatively simple sporangiophores whilst spores produced at the sexual stage of the life cycle are frequently found on very complex fruiting bodies. In each case there is an attempt to maximize the spore-bearing surface with the minimum of vegetative tissue. This results in the ability to produce enormous numbers of spores from a very limited area. For instance, *Polyporous squamosus* can produce about eleven billion (11×10^{12}) spores from each fruiting body and as there are usually several fruiting bodies on an infected tree a single infection can result in the production of as many as one hundred billion spores each year. Further, the infection will continue for several years.

2.2   The liberation of spores

Having produced these prodigious quantities of spores the fungi have to get them away from the site of production. This can be either an active or a passive process. Passive liberation is dependent on the spores, which are very lightly attached, being blown or washed off or brushed off by passing insects or higher animals. Some fungi produce materials making them attractive to particular animals or insects which come and feed on the fungus carrying away spores either on their surface or in their gut, later to be discharged in the faeces.

Active liberation of spores can involve either activity on the part of the spore, as in the swimming of zoospores, or activity of the spore support apparatus, the spore playing a passive part. The simplest means by which the support apparatus can discharge spores is the uptake of water by a mucilaginous secretion which swells and forces the spores out of the fruiting structure. In some species the mucilage may form threads several millimetres long. In a few of them the extruded spores will germinate to produce sporophores bearing wind-liberated spores. An example of this is *Gymnosporangiun juniperi-virginianae* which causes apple rust gall of red cedar. This species produces mucilaginous extrusions two or three centimetres long containing a large number of teliospores which germinate to produce sporidia and wind-dispersed spores.

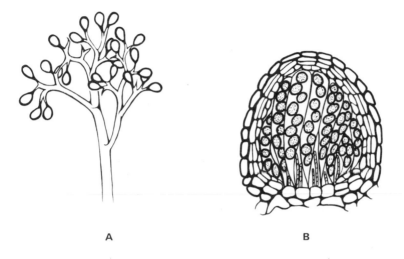

A   B

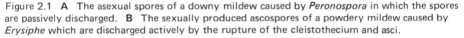

Figure 2.1   **A**   The asexual spores of a downy mildew caused by *Peronospora* in which the spores are passively discharged.   **B**   The sexually produced ascospores of a powdery mildew caused by *Erysiphe* which are discharged actively by the rupture of the cleistothecium and asci.

Some of the phycomycetes have a more violent method of spore liberation. They use an explosive mechanism which shoots the sporangia containing spores off the sporangiophore. Although not a plant pathogen the caprophilous fungus *Pilobolus* uses this method most dramatically shooting groups of spores to heights of up to two metres. Some ascomycetes also use an explosive mechanism for the liberation of spores. Pressure builds up in the sausage-shaped ascus until it ruptures at the tip, the weakest point, and the spores are shot out. As the asci are enclosed in a fruiting body with only a relatively small opening the ascus must be accurately aligned with this opening for the discharge to be successful. This alignment is a phototropic response to the light entering through the opening.

Most of the basidiomycetes have an active method of spore discharge. In the more advanced forms the basidia bearing the spores are produced in fruiting bodies either on the folded gill surface of the gill-bearing fungi or in the pores of the poly-pore fungi. In either case the spores must be shot off hard enough so that they do not lodge on the basidia lower down the gill or pore, but not so hard that they strike the opposite side of the gill or pore and lodge there. The gill or pore must be aligned absolutely vertically or the majority of the spores would be trapped. One explanation of the way this is achieved is that the spore develops on the sterigma at the tip of the basidium and when it is mature a droplet of liquid is secreted at the tip of the spore where it is attached. As this droplet increases in size, the angle

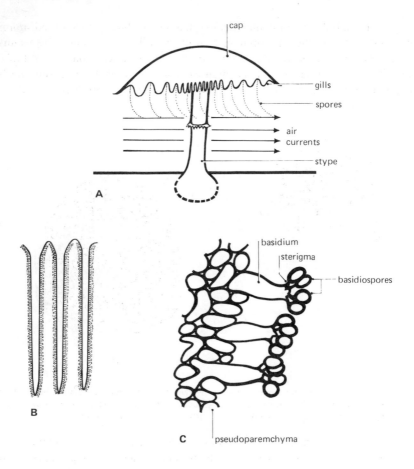

Figure 2.2 **A** Support structure for the sexually produced basidiospores of a gill-bearing fungus. **B** The gills lined with basidia. **C** The basidia bearing basidiospores.

between the surface of the liquid and the spore becomes less until surface tension is overcome and the droplet wets the spore, moving round onto one side of it. The mass of the droplet is relatively large compared to that of the spore and therefore the movement of the droplet causes a movement of the spore in the opposite direction, pushing it against the sterigma. The sudden stopping of the droplet as it sticks to the side of the spore tends to carry the spore forward away from the sterigma and this coupled with the rebound of the sterigma is sufficient to carry the spore away from the basidium and out into the space between the gills from where it falls under the influence of gravity.

Active discharge tends to be dependent on high humidities in some case free water is necessary but passive release is far less affected by humidity and is more temperature dependent. This means that active and passive discharges tend to take place at different times of day, active discharge occurring in the humid air of early morning and passive discharge in the warmth of the afternoon. The number of

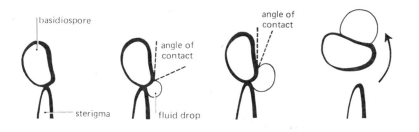

Figure 2.3 One suggested mechanism for the discharge of basidiospores. A droplet of fluid is produced at the junction of basidiospores and sterigma. The droplet enlarges until the decreasing angle of contact between it and the spore causes it to run round onto the spore, the inertia of which is enough to flick the spore free from the sterigma. (after Buller, 1922)

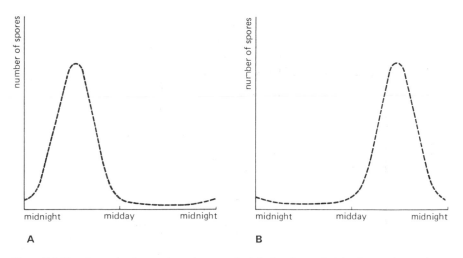

Figure 2.4 The change in the number of spores shed during day and night due to changes in humidity and temperature. **A** Damp air sporing types. **B** Dry air sporing types.

spores released varies with the time of year as well as the time of day. The number of spores in the air builds up during the summer, reaching a peak in the autumn and then decreasing in the early winter. Factors other than temperature and humidity can also affect spore discharge. Light can play an important role, triggering the discharge of spores by some fungi and inhibiting their release in others. In a few species spore discharge is triggered by the light but there is a delay of ten to twelve hours so that the spores are actually shed in the dark. Spores are discharged in some other fungi in response to an endogenous rhythm which is unaffected by external stimuli.

Air-borne spores are measured in a number of ways but they nearly all involve the impacting of spores onto a sticky surface. In the Hirst spore trap the spores are impacted onto a greased slide or strip of film drown past a narrow slit through which air and spores are drawn. This means that laid out along the side will be a record of the number of spores trapped at the different times of day. However,

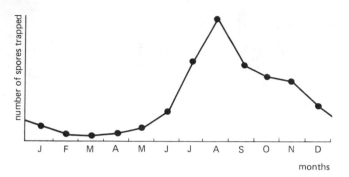

Figure 2.5 Numbers of airborne spores throughout the year, increasing during the summer and then declining again in late summer. The secondary peak in late autumn is due largely to spores from gill bearing fungi.

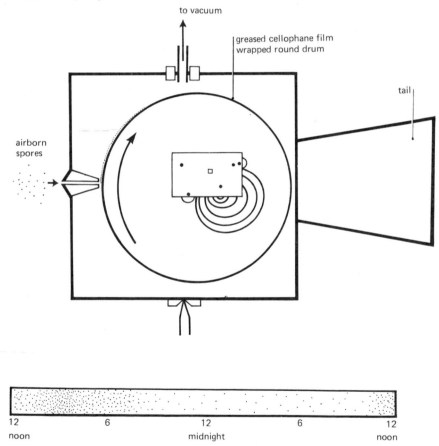

Figure 2.6 The Hirst spore trap. Air is drawn through a slit into a sealed chamber by means of a vacuum pump. The spores are carried in with the air and impacted on a greased slide or strip of cellophane film carried past the slit at a known speed by an electric or clockwork motor. The chamber is pivoted and is turned into wind by the tail. Below is a typical record of spores impacted.

many fungi cannot be identified from their spores alone and it is difficult to culture from the tightly packed mass of spores on the slide. The Bourdillon slit sampler overcomes this problem by placing a rotating petri dish of a suitable nutrient agar under a slit through which air and spores are drawn. The spores must be sufficiently separated so that they do not interfere with the growth of one another. This means that if there are a significant number of spores in the air sampling can only take place for a short period of time on each dish. The Anderson sampler is interesting in that it separates small light spores from large heavy spores. The heavy spores are impacted on the first petri dish but the lighter spores are swept past in the air stream and become impacted on the later petri dish. Although this aids in the separation of different spore types it does mean that the sample is taken at one particular interval in time. A very neat little device which does not require a vacuum pump consists of a pair of rotating greased rods on which the spores become impacted. The volume of air swept by the rods can be calculated from the

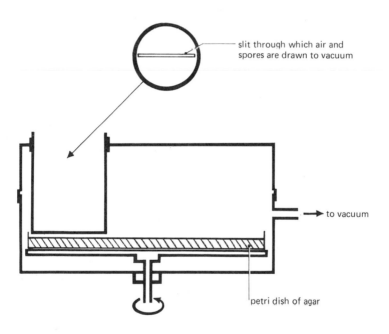

slit through which air and spores are drawn to vacuum

to vacuum

petri dish of agar

Figure 2.7 Bourdillon slit sampler. Spores are drawn in by a current of air and are impacted onto a revolving petri dish of agar containing suitable nutrients. The spores are allowed to germinate and the resulting colonies identified.

area of the face of the rods, the revolutions per minute and the time for which they were revolved. Although this also only gives the number of spores at one interval of time the small size of the rods make it easy for a number of samples to be collected and as the rods can be rotated by a clockwork motor it is very suitable for field use.

The problem with all these methods is that it is difficult to detect the obligate parasites if they do not have distinctive spores, because by their very nature they can not be grown in culture even on the most complete and complex medium.

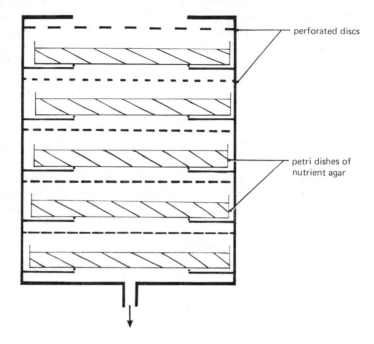

Figure 2.8 Anderson cascade spore sampler. Air is drawn through a cascade of perforated discs, the holes becoming progressively smaller so that the air velocity becomes progressively greater. Only the most easily impacted spores will be deposited on the first petri dish but as the air velocity increases spores more difficult to impact will be deposited. In this way a rough sorting of the spores will take place on the basis of their size and weight.

2.3   Dispersal by wind and water

Aerial dispersal is one of the most important ways in which spores are disseminated and although an appreciable wind is important in the long distance dispersal of spores it is not so important as at first sight it might appear. Spores are usually so small and light that they are transported by the slightest air movements and dissemination is successful even in the calmest weather.

It might be expected that the speed of fall of a spore in air would be an important factor when considering how far it will be carried before it reaches the ground. This idea was tested and the following results found:

Table 2.1   *The speed of fall of the spore in two species of fungi.*

Fungus	*Rate of fall in still air (cm/sec)*
Tilletia caries (Smut)	1·4
Bovista plumbea	0·24

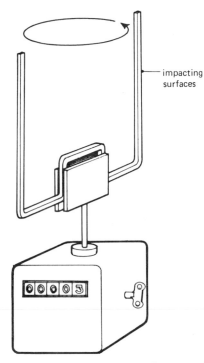

Figure 2.9 A 'roto rod' spore sampler in which instead of moving the spores to the impacting surface in a current of air, the surface is rotated so that it strikes the spore. The volume of air swept can be calculated from the frontal area of the rod, the diameter through which it is turned and the number of revolutions for which it is run. As no vacuum system is required it is very suitable for field sampling and a large number of sterile rods can be carried.

The spores were liberated together and then trapped at different distances from the source:

| | *Number of spores trapped at* | | | |
	5 m	*10 m*	*20 m*	*40 m*
T. caries	74	18	7·4	1·0
B. plumbea	74	20	5·0	1·0

(after Stephanov, 1935)

It can be seen that the number of spores trapped decreased rapidly with increasing distance from the source but the difference in rate of fall of the spores had little effect This is because the rate of fall of both of them would be slow compared to the speed of up currents and eddies and this is likely to be the case for most fungal spores.

The number of spores in a given volume of air decreases as one goes away from the source because the spores spread out sideways and upwards as they are carried away by air currents. A contour map can be drawn of spore concentrations and as expected the spore numbers decrease rapidly upwind of the source and more slowly downwind.

Spores have been detected at very great altitudes, although the majority of them are found within the first hundred metres. Spores of several common moulds were found by 'Explorer II' between 11 000 m and 22 000 m (the heights between which the trap operated) and spores of *Puccinia graminis* have been found at 4240 m over infected grain fields. It has been calculated that spores of *Alternaria* would travel nearly 4800 km before they reached the earth from a height of 1500 m, and that is ignoring the effect of any up-currents. Spores of *Puccinia graminis* would have travelled about 1200 km. Both these calculations assume the rather low average wind velocity of 20 mph at all heights. This is clearly a hypothetical calculation as the previous experimental results have shown the importance of up-currents, but it does illustrate the potentially enormous distance that spores can be carried and the importance of down-draughts and wash out by rain in bringing them to earth.

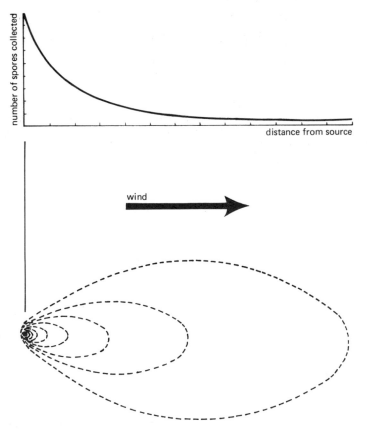

Figure 2.10 The fall-off in spores away from the source shown at the top as a graph and at the bottom as a contour map.

A very clear example of this sort of rapid wind dispersal is the epidemic of wheat stem rust which occurred in the USA in 1923. The infected plants produce spores in very large numbers which infect other plants. This in turn results in the production of even more spores and this is reflected in the accelerating rate of spread.

Spread by wind may not be as simple as this example. It is often modified by local conditions as in the case of white pine blister rust (caused by the basidiomycete *Cronartium ribicola*) which infected both the pines at the top of a slope and the currant bushes (*Ribes*) at the bottom. The question was 'Where was the source of infection?' As the pines were upwind of the currant bushes it appeared that they were the source of infection. However a detailed study showed that the spread of spores was by vertical air currents as well as horizontal ones. At night the heat retained by the trees and lake caused convection currents which carried spores from the currant bushes high into the air, from where they were carried to the pines by upper-air currents moving in the opposite direction to the low-level prevailing wind. The low-level wind caused by cold air sinking down the sides of the

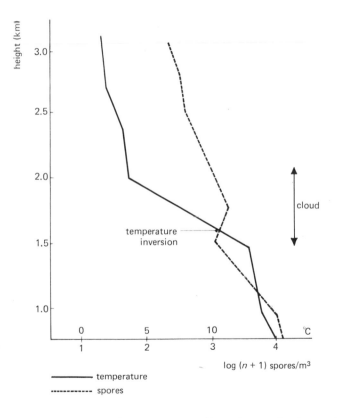

Figure 2.11 Vertical distribution of spores in the air. Note: the scale for the number of spores is logarithmic, i.e. there are 100 times more spores at ground level in this instance than at a height of 2000 metres.
(Hirst *et al*, 1967)

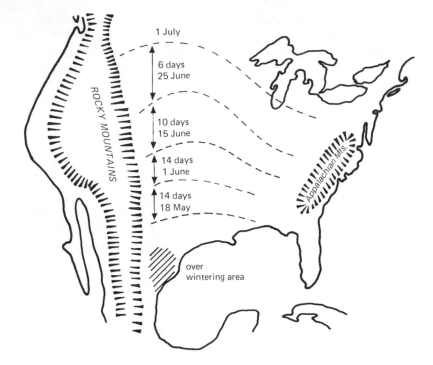

Figure 2.12 Black stem rust of wheat in N. America in the 1923 epidemic. Note the increasing rate of spread.
(United States Department of Agriculture and Minnesota Agricultural Experimental Station)

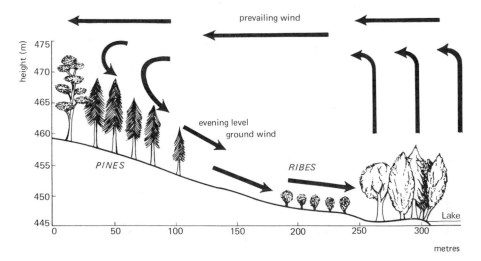

Figure 2.13 An example of vertical circulation of air causing infection to spread into the wind prevailing at ground level. In this example white pine blister spores are carried both from the pines to the *Ribes* bushes and from the *Ribes* to the pines at different heights.
(modified from Van Arsdel, 1965)

valley could of course also carry the basidiospores of *Cronartium*. This clearly demonstrates that although infection would appear to pass only from the pines to the currant bushes, in fact a two-way infection process was taking place.

Water as an agent for long distance transport of disease is relatively unimportant compared to wind. Certainly appreciable numbers of spores get carried into streams and rivers but it is relatively rare that from there they are able to get to a suitable host. This of course excludes disease agents of the aquatic environment itself. One of the ways that terrestrial pathogens can get from water to a suitable host is during irrigation and as this increases so the importance of water-borne pathogens will increase. Pathogens already shown to be spread at least in part by irrigation water include the bacterium causing angular leaf spot of cotton and *Plasmodium brassicae*, the cause of club root of *Brassicas*. Rain is very important in bringing spores to earth and it is also important in the short distance transport of infection from one plant to its neighbour by splash; a raindrop hits an infected plant and breaks up, spraying the surrounding plants with droplets which may carry spores or bacteria causing infection. Fire blight (*Erwinia amylovora*) and black rot of cabbage (*Xanthomonas campestris*) are spread between neighbouring plants in this way. This importance of rain becomes apparent when one considers the number of fungi which produce spores in association with mucilage. The mucilaginous spores will require water to separate them one from another but will become firmly attached to the host as the mucilage dries. Clearly water is most important in the dispersal and deposition of such fungal spores and bacteria.

2.4 Deposition

Deposition of air-borne spores is more difficult than one might imagine. Around a stationary object in moving air or a moving object in stationary air there is a boundary layer which will tend to sweep the spores away. Using cylinders in

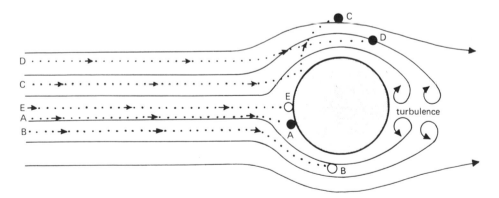

Figure 2.14 The path of heavy and light spores approaching a cylindrical object on different courses.
● Heavy spores; ○ Light spore; **A** Heavy spore impacting; **B** Light spore swept past;
C Heavy spore bouncing off; **D** Heavy spore swept past; **E** Light spore impacting. (after Ingold, 1971)

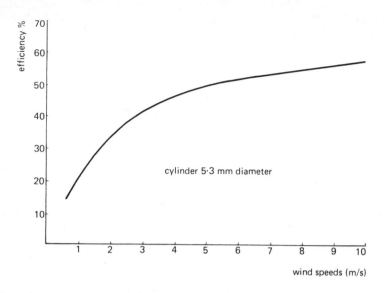

Figure 2.15 The impaction of spores on a cylindrical object at different wind speeds. Efficiency = Number of spores actually impacted divided by the number expected calculated on spore density in air stream, wind speed and projected area of target. (based on data from Gregory, 1961)

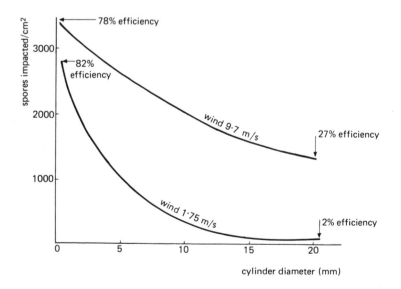

Figure 2.16 The impaction of spores on cylindrical objects of different diameters. (based on data from Gregory, 1961)

different wind speeds it has been found that impactions per unit area go down with increasing size of object and that impactions go up with increasing wind speed i.e. speed of spore.

Calculations such as these are of importance not only when considering the landing of a spore on the host but also when considering wash out of spores by rain. Large raindrops will have a large boundary layer and a low efficiency because of their large size but they will be falling very fast so that this will increase their efficiency. Also surface tension will have less effect on a large drop so that its rapid passage through the air will deform it to a more streamlined shape reducing the boundary layer but also reducing its projected area. A small raindrop on the other hand will be nearly spherical, moving slowly with a small boundary layer. Calculations can therefore be made as to what type of rain is going to cause the maximum wash out of spores of different shapes and sizes and this can be used to help in disease prediction.

The shape and size of the spore itself is important in impaction. As the size of spore increases so does its inertia; it therefore becomes less easily deflected by the boundary layers, and its efficiency of impaction increases. The ideal shape of a spore for maximum efficiency of impaction is interestingly a sphere. A long, thin streamlined shape will only be streamlined in one orientation and as the direction of air flow changes in the boundary layer the spore will present an unstreamlined projection and will be more easily swept aside than a sphere.

Experiments on deposition of spores on an actual plant, rather than a simplified model were substantially as predicted. The greatest efficiency of impaction was on the slender leaf petiole, the efficiency being lower on the stem and lowest on the leaf lamina.

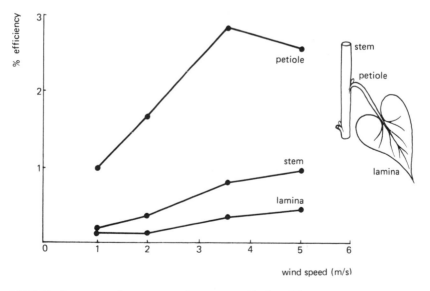

Figure 2.17 The impaction of spores on a plant stem and leaf at different wind speeds. (based on Carter, 1965)

2.5   Dispersal by insects and other animals

Dispersal by insects tends to form a different pattern from dispersal by wind, the vectors moving from plant to plant forming a radial spread pattern. This sort of pattern is especially true of soil-inhabiting vectors and wingless vectors above ground. Winged vectors are more affected by wind and may show radial spread distorted by the prevailing wind.

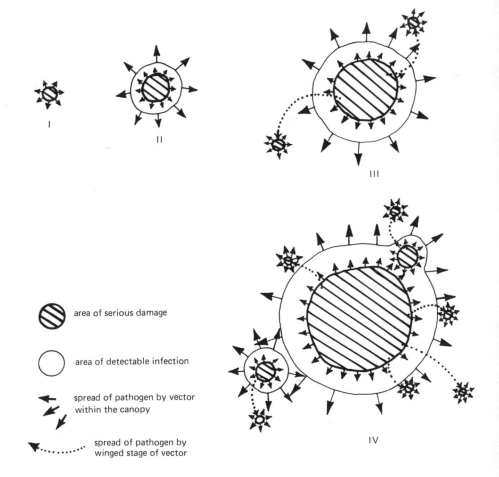

- area of serious damage
- area of detectable infection
- spread of pathogen by vector within the canopy
- spread of pathogen by winged stage of vector

Figure 2.18 The spread of disease by vectors showing the radial spread and the new foci originating from the first infection, with some of the new foci becoming incorporated into the expanding original infection.

Spread by vectors is of particular importance where the host plants are not growing close together but are separated by large numbers of unsuitable hosts. If the spores of the pathogen were wind-borne they would be spread over a wide area in which most of the species are resistant to attack. If however the spores are carried by a vector feeding specifically or even largely on one host species the likelihood of the spores being carried to a susceptible host are far greater. Often the vector will inoculate the host whilst feeding, further increasing the likelihood of successful infection.

As previously mentioned some fungi actively attract insects by secreting substances for them to feed on. The ergots of *Claviceps purpurea* for example bear conidia and at the same time produce 'stinking honey dew' which contains up to two molar sucrose. *Puccinia graminis* similarly secretes a sugary 'nectar' along with the pycniospores but having visited the fungus there is then no certainty that the insect bearing the spores will visit a susceptible host at a stage suitable for infection. A few fungi use the insect attractant powers of the host flowers. For example, *Ustilago violacea* which causes smut of campions spreads throughout the host but the production of sporangia is limited to the anthers where it replaces the developing pollen. The spores are therefore spread by the insects normally pollinating the flowers. Similarly *Botrytis anthophila* the cause of another mould in red clover is also spread by the pollinating insects.

A number of fungi are associated with boring insects, producing fruiting bodies in their galleries and being spread by the insect. An example of particular importance at the moment is dutch elm disease caused by *Ceratocystis ulmi* and carried by species of the beetle *Scolytus*. There is a close relationship between this fungus and its vector. The beetle feeds on the young twigs of elm trees and infects them with the fungus. The tree starts to die back from the infected tips and the resulting dead branches later form a suitable site for other *Scolytus* beetles to lay their eggs, which hatch and feed on the dead wood. The larvae pupate and later emerge as the adult bearing spores from the infected wood. These beetles then infect further trees during their feeding.

Occasionally a vector does not accidentally carry the pathogen but carries it deliberately. The wood wasp *Sirex gigas* attacks conifers by driving its ovipositer into the tree trunk and depositing eggs in the wood. As the egg passes down the ovipositer an oidial inoculum of *Steriem sanguinolentum* is exuded on to it from two pouches at the hind end of the wasp's abdomen. The advantage to the wasp is not clear: perhaps the larvae feed on the partly degraded wood or perhaps the fungus provides nutrients necessary for the larvae's development.

Spread by soil-inhabiting animals such as insects, nematodes oligochaetes and arthropods is often affected by barriers within the soil which may not be apparent at the surface. Such barriers would include outcrops of rock or clay which nearly reach the surface. Also regions of waterlogging and changes of soil type can form barriers that are not immediately obvious. Ditches and dense rooted hedges can likewise bar the spread of soil-borne disease by limiting the movement of the vector. These barriers can result in non-radial spread which may give the appearance of an air-borne disease.

Man is without doubt one of the most important animal vectors of plant pathogens especially over very large distances. This usually occurs in the transport of plant material such as seeds, nursery stock and timber. It is considered rare, though not unknown, for infection to be carried on the person of travellers and on manufactured goods.

Other higher animals are in some instances important disease vectors. *Plasmodiophora brassicae*, the cause of club root, can be spread if infected material, such as turnips, is fed to cattle. The spores are voided unharmed in the dung, often to infect another crop. Birds too have been suspected of transporting fungal spores, particularly where isolated outbreaks have occurred some distance from a source of infection. *Endothia parasitica*, which causes chestnut blight and has practically destroyed the entire chestnut forests of eastern USA, is thought to spread by birds, woodpeckers and treecreepers being particularly suspect. In one example birds shot around infected trees were tested for the presence of the pathogen. Two downy woodpeckers carried about half a million spores each and a brown treecreeper had over a quarter of a million spores of *Endothia* on its plumage.

Animals other than birds have also been shown to be capable of carrying infection on their coats. When a dog was run through a field of potatoes infected with potato virus X and then through a healthy crop the passage of the dog through the healthy crop was marked by the appearance of disease. It is therefore likely that other mammals can and do carry plant diseases of various sorts on their coats.

2.6   Virus transmission

Viruses are rather different from other pathogens in that they have no metabolism of their own. They are entirely dependent on the host metabolism for the synthesis of new viral particles. These particles are the dispersive phase of the life cycle and the only stage at which they can be readily distinguished from their host as a separate entity. All plant viruses lack even the simplest dispersal mechanism, being totally dependent on outside transport mechanisms. The agents involved in their transmission are thus of particular interest to us.

Viruses can be divided into two groups, vector carried and non-vector carried. Each group can then be sub-divided in to soil-borne and air-borne. The non-vector transmission of viruses in infected soil and plant material is similar to the transmission of other diseases, so we will concern ourselves with vector transmission.

Surprisingly one group of agents involved in the transmission of viruses in the soil is fungi. The fungus is almost always a chytrid, frequently a species of *Olpidium*. Viruses carried in this way include lettuce big vein, tobacco necrosis and a number of cereal viruses. When plants are watered with a suspension of these viruses no infection results, but if infected roots are incorporated into the soil the plants rapidly become infected. The involvement of the fungus was shown in a number of ways, the most dramatic being the halt brought to virus infection by the application of a fungicide to the soil. Another agent important in the spread of viruses in the soil is eelworms. Again the application of a virus suspension results in no infection if the plants are growing in sterilized soil. In unsterile soil, infection frequently occurs and the hand picking of eelworms from around infected roots and putting them onto roots in sterile soil always results in virus infection. Transmission of the virus can again be stopped by killing the vector, in this case with a nematocide. There is a wide range of viruses transmitted by nematodes and the relationship is probably fairly specific between the virus and nematode.

Table 2.2 *The specificity of some viruses to their nematode vectors.*

Nematode vector	Virus
Xiphinema index	Arabis mozaic virus (grape vine strain)
	Grape vine leaf roll virus
X. diversicaudatum	Arabis mozaic virus (type strain)
X. americanus	Peach yellow bud mozaic virus
Longidorus attenuata	Tomato black ring virus
Trichodorus primitivus	Tobacco rattle virus (English)
T. pachydermis	Tobacco rattle virus (Dutch)

Air-borne vectors are by far the most important means of virus transmission. This is mainly by insects, including aphids, jassids, thrips, mealy bugs and beetles. Viruses carried by these vectors can be divided into three groups, non-persistent, semi-persistent and persistent viruses. The non-persistent viruses are those whose vectors lose their infectivity within an hour or so of feeding on an infected plant. Semi-persistent viruses cause their vector to remain infectious for several hours and vectors of persistent viruses are infectious for several days after an infection feed. The type of virus and the plant from which it originates are important when considering the period for which the vector will be infectious; surprisingly the species of vector is less critical in the longevity of the virus. Knowledge of the plant source of the virus is probably important because of the other materials such as proteases ribonucleases and organic acids taken up during the feeding of the insect which could result in the inactivation of the virus.

Vectors of non-persistent viruses are more infectious if the infection feed is short, about five minutes usually being optimum, but in some cases as little as fifteen seconds gives maximum infectivity. The same thing applies for transmission feeds with regard to subsequent infectivity, infectivity being highest after short feeds. It would seem that feeds of large amounts of plant sap reduce the infectivity of the virus. This is perhaps because the virus is in contact longer with plant cell constituents. A period of starvation between the infection feed and transmission feed, i.e. no further uptake of non-infectious sap, increases the infectivity of the vector. This makes a vector from outside a crop likely to be more infectious than one from within a crop.

Many non-persistent viruses are thought to be carried on the mouthparts of the insects and this accounts for the short infectious period and also for the loss of infectivity when feeding on healthy plants. The virus particles are simply rubbed off. Treatment of the vector's mouth parts with agents known to inactivate the viruses, such as formalin or ionizing radiation, greatly reduces or completely destroys the ability of the vector to cause infection, also strongly suggesting that the virus is carried on the mouth parts.

The vectors of semi-persistent viruses remain infectious for a considerable period and irradiation of the mouthparts has in some instances been shown not to reduce infectivity. However there is good evidence that the virus does not persist through a moult of the vector. When an aphid moults it sheds not only its outer skin but its stylets, foregut and hindgut. It therefore seems likely that the virus is carried on one of these regions.

Persistent viruses always have a latent period between the infection feed and the transmission feed, that is a period in which they are non-infectious. Very few persistent viruses can be mechanically transmitted in the sap from an infected plant. These two facts could be explained by the hypothesis that there is a very low concentration of virus particles in the sap of an infected plant, too low to cause infection in another plant, and these virus particles multiply up to an infectious concentration in the vector during the latent period. This would mean the virus is capable of instructing such different types of cell as those found in the host plant and insect vector, how to synthesize new viral particles! There is good evidence for this hypothesis, since some viruses have been shown to be present in the salivary gland of the vector and to be serially transmitted from aphid to aphid by injections of haemolymph. It is thought that the virus infection reduces the life span of the aphid and the leaf hopper. There is visible damage to the fat bodies of the infected hoppers. The cytoplasm of the cells becomes reticulate with large vacuoles, the nucleus becomes stellate and some cells are completely destroyed. It has been demonstrated that some viruses can be transmitted through the eggs of the vector over a great many generations. As the vectors were fed on resistant host plants there must have been multiplication of the virus particles within the vector for the enormous number of offspring produced over a number of generations to have remained infectious.

It has now been reported that clover phyllody disease and some of the yellows diseases are in fact caused by mycoplasma rather than by viruses as previously thought. Mycoplasma are very tiny organisms much smaller than most bacteria and lacking any rigid cell wall. They are nutritionally very fastidious, normally requiring the presence of living cells, but unlike viruses they can be cultured under suitable conditions on extremely rich artificial media. The aetiology of mycoplasma diseases is very similar to that of persistent viruses and it is possible that some of the diseases we now consider to be due to viruses may in fact be caused by mycoplasma-like organisms. There is however good evidence that not all persistent virus diseases are due to such organisms.

2.7 Vector activity

Vector activity is important in the way a disease spreads, whether it is caused by a fungus, a bacterium or a virus. It has been found under laboratory conditions that *Aphis ramni* is as effective as a vector of potato leaf roll virus as *Myzus persicae.* In the field however it is not nearly so effective because *A. ramni* is sluggish, moving very little and feeding mainly from one plant, whereas *M. persicae* is active, moving from one plant to another spreading the virus. The age of the vector can also be important. It has been shown that older, winged aphids are less active than younger adults due to atrophication of the wing muscles with age and that this process was triggered by reproduction.

If the host is highly suitable for the vector fewer probes are made before the aphid settles down and starts steady feeding. It is therefore probable that a plant suitable for the aphid will be inoculated with a smaller amount of virus than a less suitable plant. It has also been observed that young aphids make more probes than older ones before continuous feeding, which will make younger aphids more infectious than older ones.

The site of feeding of the vector is important. Cauliflower is infected with both cabbage black ring spot virus (CBRSV) and cauliflower mozaic virus (CMV). The vector of both being the aphid *Myzus persicae*. Despite the fact that CBRSV is present in higher concentrations in the cauliflower and has a wider host range than CMV it is less infectious. This is because CBRSV is present in high concentrations in the older cauliflower leaves but in low concentrations in the younger leaves where the aphids feed. Similarly the age of a crop will affect its susceptibility to infection by the vector. The age of sugar beet is important in its infection with sugar beet yellows.

Table 2.3 *How the percentage of sugar beet plants infected depends on their age at peak activity of the vector.*

Date of planting	Percentage of plants infected when mature
1 April	35
15 April	53
1 May	70

The early sowing results in the plants being fairly mature at the time the vector is active and the more mature plants are less attractive to the vector than the younger plants.

The spacing between plants can also be important. Aphids tend to fly round a dense stand of plants and only alight at its edges, so that infection proceeds from the edge. In an open stand aphids land in a random manner over the entire crop causing the infection to proceed more rapidly. Dense planting however tends to increase the chance of spread within a crop either by contact between plants or by crawling insects. The suitability of landing sites can affect the spread of air-borne vectors. Aphids tend to land in sheltered conditions, so that infections frequently start from a region sheltered from the prevailing wind by hedges or buildings.

2.8 Seed-borne infection

The importance of seed-borne diseases is obvious: it not only causes the early infection of that crop but can often form the centre of a disease outbreak spreading to neighbouring crops. Because of the transport of seed over long distances it can be the means of introducing a disease in areas previously free of it. It also enables pathogens to survive from one season to another.

Seed contaminated on the outside with the spores of a pathogen, such as bunt, can be treated to remove or kill the pathogen provided the seed is known to be contaminated. Treatment with seed dressings is of great benefit in controlling such diseases and it is now fairly routine to use seed dressings on all agricultural seed in case it is contaminated with pathogenic spores and to decrease the likelihood of the germinating seed becoming infected with soil-borne pathogens.

Seed dressings will not however control infections within the seed. The grass *Lolium* for instance is almost always infected with an endophytic fungus which is carried through the ovule to the next generation The loose smut (*Ustilago*) infects the flowers of cereals and enters a resting stage in the seed. When the seed germinates the fungus grows up within it to produce a mass of spores in the flower spike which will infect neighbouring plants. Fortunately, this fungus is more sensitive to heat than the grain it infects and heating to $50-55°C$ for fifteen minutes destroys most of the fungus, but control of seed-borne diseases is rarely this easy.

Grasses are not the only plants to carry pathogens in and on their seeds. The anthracnose of french and runner beans caused by *Colletotrichum lindemuthianum* is carried on the seed. The fungus attacks all parts of the bean plant producing local lesions of disintegrating tissue in the centre of which small mucilaginous conidia are produced. Infections on the pod will often penetrate right through to the seed which becomes infected with lesions on the cotyledons. This carries the infection to the next generation.

Vegetative propagation is a serious source of infection of fungal, bacterial and viral diseases. Many plant diseases are not transmitted through the seed, so that seed from plants infected with such diseases will give rise to healthy seedlings. With vegetative propagation it is very rare to be able to obtain disease free material from infected plants. Culturing the apical meristems sometimes enables healthy plants to be clonally propagated from infected material because the meristematic cells may be dividing sufficiently fast for their growth to be greater than that of the pathogen and they thus remain disease free. Heat treatment of infected material can in some instances rid plant material of its pathogen but this depends on the pathogen being more heat sensitive than the host. Systemic fungicides, bacterio-cides and antibiotics are now also being used to rid particularly valuable clonal material of disease but frequently it is not possible to maintain vegetative plant material free of infection and the vigour of a clone will gradually decline as the level of infection rises.

Virus transmission is common in both vegetative material and seeds. There is a great variation in the percentage of the seed from a diseased plant carrying infection, depending on the growth rate of the host, time of infection, site of infection etc. In lettuce for example, the percentage of infected seed from a virus-infected plant varies between 0·2 per cent and 14 per cent. In fact if the plant is infected after flowering it is rare for any of the seed to be found to carry the virus. Seed-borne virus infections are found in all families of plants and it is probable that it is far more common than is generally realized.

Chapter 3
Epidemiology

3.1 What is epidemiology?

Epidemiology is a study of populations. The fate of an individual plant is immaterial except in so far as it affects the population, e.g. forms a centre of infection.

In the introduction we have seen how diseases can start from small beginnings and spread to pandemic proportions. On the other occasions not reported in history, diseases can start from small beginnings and die away completely, or spread very slowly. Let us therefore investigate the reasons for the varying rate of spread of a disease. The shorthand of mathematics is useful in such an investigation in that it enables us to see how the variables we are dealing with interact.

Clearly many factors will affect the course of an epidemic but these can be put under three headings: (i) original amount of inoculum; (ii) rate of progress of the disease in the population; (iii) time during which the disease can progress. This can be seen if a graph is drawn of the percentage infection against time.

Percentage infection will vary with the type of disease and the method of recording it. It might simply be the percentage of plants in a population infected or it could be the percentage leaf area killed or the percentage of ears of grain destroyed.

In nearly all instances however the course of the disease will follow a sigmoid curve, exponential or logarithmic at first, flattening out later. This curve is obvious when you think about it. The pathogen produces spores or 'propagules' of some sort which spread the disease. They will of course produce more 'propagules' than were required to produce themselves. For example, one spore will cause an infection producing millions of spores, each one of which is capable of producing an infection that will yield further millions of spores. This results in the exponential increase at the beginning of the sigmoid curve. The rate of spread of the disease within the population will fall off as the number of plants available for infection decreases i.e. a decrease in the number of plants not already infected. The same is of course true if one is measuring infected leaf area or ears of grain etc; infection can never exceed 100 per cent unless a plant recovers and is re-infected. The time scale will vary enormously from one disease to another, some measured in days or even hours and others in years.

A disease may not follow quite this pattern if for example sporulation occurs at regular times of the year. This would mean that growth of the disease during the

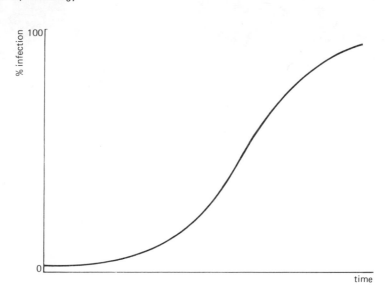

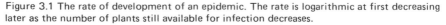

Figure 3.1 The rate of development of an epidemic. The rate is logarithmic at first decreasing later as the number of plants still available for infection decreases.

year would be vegetative and relatively slow and then suddenly there would be a rapid increase in the disease following sporulation. Taken over a long period of time however, even a disease like this would have a sigmoid growth curve but it would be a slightly 'bumpy' curve.

The exponential part at the beginning of a disease growth curve will have the same mathematical formula as compound interest.

$$x = x_O \, e^{rt}$$

where; x = proportion of disease at this instant of time
x_O = original proportion of disease i.e. inoculum
e = mathematical constant e
r = rate of increase of disease
t = time

Let us look at the case of a bacterial disease where the bacterium will divide into two in unit time and these will divide into four in the next unit time and eight in the next then sixteen, thirty-two, sixty-four etc. Here the rate r will depend on the doubling time. In the case of fungal diseases one spore may grow and produce many sporangia each bearing thousands of spores i.e. each spore produces millions of spores in unit time rather than merely doubling in number. Fortunately the time taken for the fungal spore to produce more spores is very much greater than the time taken for a bacterium to double. The rate of increase of a fungal disease r can however still be very high. If a bacterial infection is such that the bacterium doubles every three hours at the end of six days there would be of the order of $2 \cdot 8 \times 10^{14}$ bacteria. If a fungal infection takes three days to go from a single spore to the production of new spores and it produces ten million spores it will have produced 10^{14} spores at the end of six days. It can be seen that despite the different generation time the rate r of the diseases is quite comparable.

From the compound interest formula it is easy to see that reduction in any one of the factors mentioned as affecting the course of an epidemic (original amount of inoculum, rate of progress of disease, time) will cause a reduction in the incidence of disease in the population but these factors will not alter the incidence of disease to the same extent or in the same way.

3.2 Using $x = x_O e^{rt}$ in disease control

Anything which reduces x_O, e, r or t will reduce the amount of disease present, x. As e is a mathematical constant only the other three factors can be altered.

The most obvious way of reducing x_O (the disease inoculum) is the removal or sterilization of infected plant material. There are however other ways. One of these is resistance of the host. Frequently there are a number of strains or races of the same species of pathogen and it is sometimes possible to breed a plant that is resistant to one or more of these races. If a plant is resistant to 99 per cent of the spores of a pathogen falling on it this will have the same effect on the course of the disease as removal of 99 per cent of the source of the spores.

This sort of resistance is sometimes called 'vertical resistance' because it moves the start of the sigmoid growth curve down the vertical axis. This is one of the commonest forms of resistance used in plant breeding, but it does leave the crop very liable to infection from a race to which it is not resistant.

Methods of reducing x_O other than breeding for resistance are commonly bracketed together as 'sanitation'. This is largely the elimination or reduction in the amount of the original inoculum. Other methods of reducing the effectiveness of the inoculum can also be considered as sanitation and have the same epidemiological effect.

Some diseases are localized so that import restrictions and quarantine measures are effective methods of stopping their spread. An example of this is coffee rust which has so far been excluded from the western hemisphere.

Seed is often infected with the spores of pathogens and a number of forms of sanitation are used to control these. Some pathogens can be destroyed by chemical treatment of the seed (seed dressings) whereas others may be resistant to chemical treatment and only disease-free seed used to produce healthy plants. This is ensured by rigorous inspection and the culturing of seed samples to detect the pathogen if it is present. Yet another method is the use of seed from regions where the disease in question does not occur.

Vegetative propagation causes the spread of many diseases unless great care is taken to maintain disease-free clones. For example potatoe tubers can carry a wide range of bacterial, fungal and virus pathogens and the planting of certified disease-free 'seed' potatoes is practised as part of the means of reducing the level of inoculum.

As many pathogens attack specific hosts and die in the absence of that host, crop rotation is a form of sanitation. The pathogen slowly dies in the soil during the period in which crops are grown that it is unable to parasitize. Some fungal spores and sclerotia only germinate near the surface so that deep ploughing reduces the incidence of these diseases, eg *Sclerotium rolfsii*. More obvious methods are also used to reduce the level of inoculum. Clearing leaves from orchards or sterilizing them with chemicals reduces the incidence of apple scab by stopping the fungus overwintering in the fallen leaves.

Having planted healthy seeds or cuttings it is of course vital to ensure they do not become infected in the nursery. Compost and seed trays are sterilized by heat or chemicals and even the the nursery beds themselves are often sterilized to reduce infection. All these and many more methods are used to reduce the level of inoculum at the beginning of the growing season.

Anything that reduces the rate of progress of the disease will reduce r. Chemical treatment with fungicides that slow the growth of the fungus or by killing a large percentage of the spores of a pathogen landing on a host lead will slow the development of a disease i.e. reduce r.

Plant breeding is also used to reduce r. Plants are bred which are not totally resistant to the pathogen but are only attacked slowly so that the generation time for all races of the pathogen is long. This will alter the slope r by altering the time base (horizontal axis) and is therefore often called horizontal resistance.

Reducing t reduces the time available for the pathogen to multiply on the crop and damage it. So that short growing seasons reduce the incidence of disease, all other factors being equal. Reduction in the growing season can be achieved by growing rapidly maturing varieties or by speeding up growth by improved culture techniques, fertilizers etc.

3.3 The effect on an epidemic of reducing x_O

Let us take two hypothetical identical fields, identical in every respect. In fact as hypothetical fields are very cheap let us take a whole lot of them to reduce variation even further. These fields are divided into two groups, one in which sanitation in all forms is practised and the other where it is not practised in any form. Let the proportion of tissue originally infected in the unsanitized fields be x_O and in the sanitized fields be x_{OS}.

Consider the exponential (logarithmic) stage of the disease where $x = x_O e^{rt}$ is applicable:

$$x = x_O e^{rt} \quad \text{unsanitized}$$

$$x_S = x_{OS} e^{rt} \quad \text{sanitized}$$

where; x and x_S are the final amounts of disease present in the crop.

Division gives:

$$\frac{x}{x_S} = \frac{x_O}{x_{OS}}$$

i.e. the effect of sanitation remains constant throughout the exponential phase of the disease.

So if 99 per cent of the inoculum is removed by sanitation there will be at harvest time only 1 per cent of the incidence of disease that there would have been without sanitation.

As the incidence of disease is increasing exponentially with time, our previous findings can be put in another way. That is, it would take the same time for the incidence of the disease to rise from 0·1-1 per cent as from 1-10 per cent. So that sanitation is buying time, the time taken for the pathogen to multiply up to the level that it would have started at without sanitation.

Let us look at another example in which r is taken as 0·2 per unit per day (i.e. there is a 20 per cent increase in the pathogen population each day) and 80 per cent of the inoculum has been destroyed by sanitation, i.e.

$$x_O / x_{OS} = 100/(100\text{-}80) = 5.$$

$$x_O = x_{OS}e^{rt}$$

therefore $\quad \Delta t = \dfrac{1}{r} \log_e \dfrac{x_O}{x_{OS}}$

where; Δt = the delay in the course of the disease due to sanitation

now $\quad r = 0\text{·}2$ and $\dfrac{x_O}{x_{OS}} = 5$

so $\quad \Delta t = \dfrac{1}{0\text{·}2} \log_e 5$

$$= \dfrac{2\text{·}3}{0\text{·}2} \log_{10} 5$$

$$= 8 \text{ days}$$

i.e. with a 20 per cent increase in the pathogen population each day, 80 per cent sanitation delays the course of the disease by eight days. This can be expressed more clearly as a graph which shows that throughout the course of the disease the effect of sanitation is an eight day delay in the disease.

So in this instance reduction of the inoculum by 80 per cent will delay the disease by eight days. This time will of course be different for organisms with different growth rates r, but in all instances reduction of x_O will increase the time taken for a disease to reach epidemic proportions.

How is this delay in disease reflected in terms of increased yield? Let us look at the example of *Phytophthora* blight of potatoes. It has been estimated by some workers that growth of potato tubers stops when 75 per cent of the foliage has been destroyed. Workers with other varieties of potato have put this figure at nearer 40-50 per cent. Whichever figure is taken it would seem to be just about within the exponential part of the disease curve. So our previous calculations still hold good. Let us assume a rate $r = 0\text{·}46$ per unit per day and that 99 per cent of the inoculum is destroyed (such as culled potatoes etc) i.e. $x_O / x_{OS} = 100$

from the equation $\quad \Delta t = \dfrac{1}{r} \log_e \dfrac{x_O}{x_{OS}}$

$$\Delta t = 10 \text{ days}$$

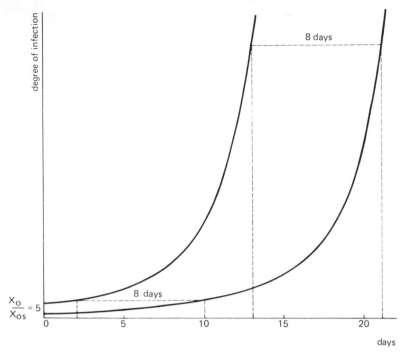

Figure 3.2 Showing that the delay time caused by a given level of sanitation is constant throughout the logarithmic phase of a disease.

From the two graphs of potato yield and percentage blight it can be seen that a delay in onset of ten days increases the yield from 75-90 per cent of the healthy yield i.e. sanitation has reduced the loss by 60 per cent. This means that sanitation has resulted in an increase in yield of something like 20 per cent.

In the formula $\Delta t = 1/r \; [\log_e(x_0/x_{0S})]$ it will be noticed that it is the logarithm of x_0/x_{0S} that affects the delay time, so that there is a decreasing effect on delay time as the sanitation ratio rises. The 'law of diminishing returns' holds good. The moral is that quick and easy sanitation yields a better return in the field than trying to eradicate the last traces of disease. In situations where there is little likelihood of infection from outside eg nurseries and glasshouses, it is frequently possible to eliminate the disease completely by sanitation and here of course it is well worth while using thorough sanitation to eliminate the pathogen totally.

These calculations only hold good for the logarithmic phase of the disease growth curve. As the levels of infection increase then the amount of crop still to be infected becomes limiting. At such time it will be $x/(1 - x)$ that will be reduced by the logarithm of the sanitation ratio, where $(1 - x)$ is the amount of the crop left to infect.

It is difficult to demonstrate in practice the effects of sanitation. This is because it is difficult to compare two fields one sanitized and the other unsanitized since for them to have similar climatic conditions and similar soil they must be near each

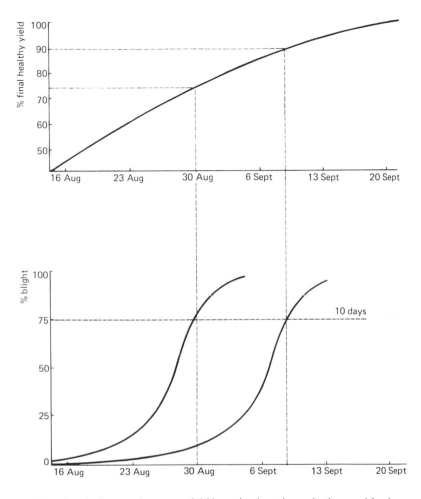

Figure 3.3 Showing the increase in potato yield brought about by sanitation resulting in a delay time of ten days.
(after Van Der Plank)

other, (ideally the two treatments should be carried out on small plots laid out in a standard experimental pattern to reduce variation of uncontrolled factors to a minimum) and the inoculum from the unsanitized field will spread to the sanitized field and destroy any meaningful comparison. Using the same field with the same crop on different years is of little help because of the climatic effect on the course of a disease.

Although difficult to demonstrate, the usefulness of sanitation is undisputed and is particularly marked in crops like cereals, far more so than in a crop like potatoes, because potatoes can be harvested as a useful crop before they have reached their maximum size. This is not so for cereals as grain can only be harvested when it is mature. Taking the rate as 0·46 per unit per day let us look at the effect of removing

90 per cent, 99 per cent and 99·9 per cent of the inoculum, eg barberry bushes in the case of black stem rust of wheat:

$$t = \frac{1}{r} \log_e \frac{x_O}{x_{OS}}$$

$$= \frac{2 \cdot 3}{r} \log_{10} \frac{x_O}{x_{OS}}$$

90 per cent sanitation

$$t = \frac{2 \cdot 3}{0 \cdot 46} \log_{10} 10$$

$$= \frac{2 \cdot 3}{0 \cdot 46} \times 1 = 5 \text{ days}$$

99 per cent sanitation

$$t = \frac{2 \cdot 3}{0 \cdot 46} \log_{10} 100$$

$$= \frac{2 \cdot 3}{0 \cdot 46} \times 2 = 10 \text{ days}$$

99·9 per cent sanitation

$$t = \frac{2 \cdot 3}{0 \cdot 46} \log_{10} 1000$$

$$= \frac{2 \cdot 3}{0 \cdot 46} \times 3 = 15 \text{ days}$$

i.e. increasing sanitation by a factor of ten delays the disease by five days.

Suppose disease at harvest is equal to 75 per cent as it was in the potato crop we considered

$$\text{that is} \qquad x = 0 \cdot 75$$

$$\text{then} \qquad \log_{10} \left[\frac{x}{1-x} \right] = \log_{10} 3 = 0 \cdot 477$$

with a sanitation ratio of 100, the log of which is equal to 2

$$\text{then} \qquad \log_{10} \left[\frac{x_S}{1-x_S} \right] = 0 \cdot 477 - 2 = \bar{2} \cdot 477$$

$$\frac{x_S}{1-x_S} \simeq 0 \cdot 03$$

$$x_S = 0 \cdot 03 - 0 \cdot 03 \, x_S$$

$$1 \cdot 03 x_S = 0 \cdot 03$$

$$x_S \simeq 0 \cdot 029 = 2 \cdot 9 \text{ per cent}$$

As few of the infected plants will be harvestable the yield will have been increased from 25 per cent to about 97 per cent, about a 400 per cent increase in yield as compared to the 20 per cent increase in yield that this degree of sanitation would have achieved in a potato crop.

In these examples we have taken the rate r of the disease as the very high figure of 0·46 per unit per day. If the lower value of 0·23 per unit per day is used how will sanitation affect the onset of the disease?
For 99 per cent sanitation

$$t = \frac{2·3}{0·23} \ \log_{10} \ 100$$

$$\frac{2·3}{0·23} \times 2 = 20 \text{ days}$$

i.e. halving the rate doubles the delay time.

These calculations of the effect of sanitation on wheat rust fit in with the field observations made on the role of the barberry bush at first subjectively and later by more scientific observations. Near Northfields, Minnesota, in the summer of 1922 there were 175 large heavily rusted bushes in one planting. By 26 May the rust had spread 30 m from the bushes. By the 6 June it had spread more than 2 500 m, by the 12 June 4000 m, by 17 June 6400 m and eventually it had spread over 16 000 m from the infected barberrry bushes. In another instance rust on wheat was being investigated and a contour map indicating the degree of infection was drawn up. From this the centre of the infection was predicted and sure enough barberry bushes were found there.

Barberry eradication is obviously useful. Unfortunately the situation is not quite as simple as this would seem to indicate. Although *Puccinia* cannot overwinter on wheat straw in the northern latitudes it can overwinter in straw in the south. When there are climatic conditions in the north that would lead to a serious spread of the disease from barberry bushes, the disease rapidly spreads in from the south. So has anything been gained by sanitation? The answer is yes, in two ways. During years when rust does not become epidemic the low background level of infection that causes a continual small loss is reduced by barberry eradication. Far more important perhaps is that the fungus can only undergo its sexual stage on the barberry bush. So that eradication of the bushes reduces the chances of genetic reassortment in the fungus and thus reduces the likelihood of the fungus becoming virulent towards strains of wheat that have been bred for resistance to the disease.

So far we have investigated delay time in the onset of a disease and the effect of this on the level of infection when the plant is mature. How much saving in damage to the plant, and thus increase in yield does this represent? This is a difficult question to answer and to make any sort of assessment we must make two assumptions: (i) Injury to the host as a whole is proportional to the amount of tissue infected, eg two pustules do twice as much damage as one pustule; (ii) injury is proportional to the duration of the disease, eg a pustule produced ten days before a plant is mature does twice as much harm as one produced five days before it is mature. Both these are reasonable and seem to be borne out on an empirical basis in experiments.

Making these assumptions we can say that the loss in yield is going to be proportional to the area under the disease progress curve (Figure 3.4). The area under such a curve can be calculated fairly simply using integral calculus. However, we are not interested in the area under the curve; what we wish to know is the ratio of the areas under progress curves of a disease with and without sanitation i.e. the reduction in damage to a crop resulting from sanitation. Happily this makes the calculation very much easier, rather than harder. Provided we assume the development of the disease to be the same in the two cases, differing only in the time of onset, most of the factors cancel out (i.e. the formula for the two curves differ only in their constants). The ration of the two areas A and B (Figure 3.5) is then found to be the same as the ratio between the two levels of infection, x and y.

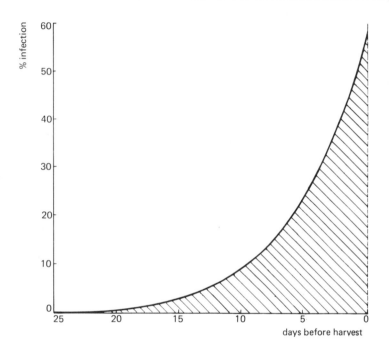

Figure 3.4 Taking the total damage to a crop as the area under the disease progress curve, the greater importance of low levels of infection early in the season compared to higher levels of infection late in the season are clearly seen.

This is true at any time during the exponential phase of the disease.

This means that for diseases with a given rate of progress the total damage is proportional to the final level of infection. This is not of course true for diseases with different rates of progress r, in which the area under the curve will no longer be proportional to the final level of infection. It is interesting to note that the area under the curve is inversely proportional to the logarithm of the delay time, i.e. as we have already discovered maximum benefit is obtained from elementary sanitation.

These calculations only hold good whilst the disease is in the exponential phase and there is no interaction between infection sites. Later in the disease when re-infection of already infected tissue occurs the rate is no longer exponential and the effect of sanitation is less. On the other hand, if death of the host occurs before there is interaction of infection sites sanitation will have a more marked effect on the yield late in the season than on the percentage reduction of infection early in the season. This is because sanitation may mean infection never reaches fatal proportions on a plant so that there is some yield as distinct from no yield on a dead plant.

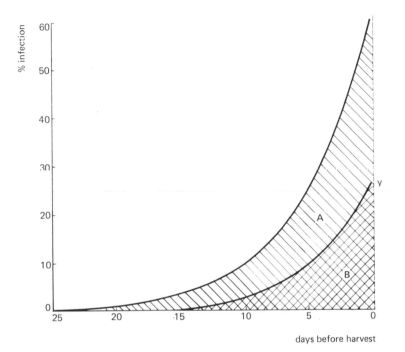

Figure 3.5 Showing the decrease in area under the disease progress curve brought about by a delay in the development of a disease. The rate of progress of the two diseases illustrated are in fact identical, the lower curve having been delayed by 10 days.

3.4 The effect on an epidemic of reducing *r*

We have seen how reduction in the original amount of inoculum will have some effect on reducing the incidence of disease. However the rate of spread *r* and the time of spread *t* are often much more important than the amount of inoculum in determining the final amount of infection and the total damage done to a crop. A high infection rate and a long period over which infection can increase are largely the factors causing sanitation to be sometimes an inadequate method of disease control.

To make it clear that the level of infection at a particular time is not the sole criteria for assessing total damage to a crop, (Figure 3.6) shows the course of two disease outbreaks, (1) a very aggressive strain of the pathogen which multiples rapidly from a low level and (2) a less aggressive strain which multiples rather more slowly from considerable inoculum. At the time of assessment the amount of

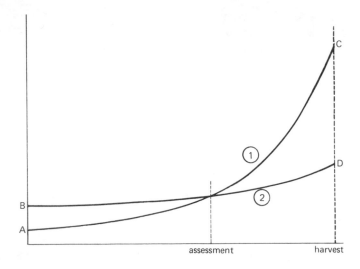

Figure 3.6 Showing the importance of the rate of development of a disease and how this cannot be assessed by a single observation. Observation at the part indicated would give the impression that disease outbreaks 1 and 2 were of the same seriousness, while observations before this would indicate outbreak 2 to be the more serious.
A = 1/3 B
C = 2·5 D
Rate 1 is 25% greater than rate 2

infection is the same in each case but prior to this point strain (1) has always had less infection than strain (2) and therefore it is likely to have done less damage than strain (2). Following this the more rapid growth of strain (1) produces higher levels of infection so that the damage done to the crop by the harvest time is considerably greater in the case of strain (1) than strain (2) despite the initial low levels of inoculum.

Let us consider a real example. Potatoes can be considered to stop producing any increase in tuber weight when 75 per cent of the foliage is destroyed. In the Netherlands a blight outbreak originating from infected tubers amongst the 'seed' was investigated. It was found that one infected tuber per square kilometre of potatoes was sufficient to initiate an epidemic under suitable climatic conditions. If there were 4×10^6 plants per square kilometre and 750 lesions per plant, there would be 3×10^9 lesions per square kilometre. If we assume that the original blighted shoot is equivalent to three lesions, this makes an increase of 10^9 lesions in a season of, say, from 1 June to 28 August, which is 88 days,

therefore $t_2 - t_1 = 88$

and $\dfrac{x_2}{x_1} = 10^9$

where; t_1 = initial time and t_2 = the final time, and x_1 = initial infection and x_2 = the final infection

Now $\qquad r = \dfrac{2 \cdot 3}{t_2 - t_1} \log_{10} \left[\dfrac{x_2}{x_1} \times \dfrac{(1 - x_1)}{(1 - x_2)} \right]$

As the initial inoculum is very small, we can take $(1 - x_1)$ is equal to 1. The final infection has destroyed 75 per cent of the foliage, therefore x_2 is equal to $0 \cdot 75$, so

$$1 - x_2 = 0 \cdot 25$$

therefore $\qquad r = \dfrac{2 \cdot 3}{88} \log_{10} \left[\dfrac{10^9}{1} \times \dfrac{1}{0 \cdot 25} \right]$

$$= \dfrac{2 \cdot 3}{88} \log_{10} 4 \times 10^9$$

$$= \dfrac{2 \cdot 3}{88} \times 9 \cdot 602$$

$$= 0 \cdot 251 \text{ per unit per day}$$

Let us suppose that there is a year with the weather slightly more favourable to the spread of blight where r' is equal to $0 \cdot 3$ per unit per day. If blight starts spreading on the same day (1 June) how much will the inoculum have to be reduced so that the increase in potato size is not halted until 28 August i.e. destruction of 75 per cent of the foliage?

$$x = x_O \, e^{rt}$$

$$x_S = x_{OS} \, e^{r' t}$$

As we are concerned with the point at which 75 per cent defoliation occurs in both instances,

$$x = x_S$$

therefore $\qquad x_O \, e^{rt} = x_{OS} \, e^{r' t}$

Initially there were 3 lesions and at the rate of $0 \cdot 251$ per unit per day it took 88 days to reach 75 per cent defoliation

$$3 \, {}_c 0 \cdot 251 \times 88 = x_{OS} \, e^{r' t}$$

We are now concerned with the new rate of $0 \cdot 3$ per unit per day for the same period,

therefore $\qquad 3 \, e^{\, 0 \cdot 251 \times 88} = x_{OS} \, e^{\, 0 \cdot 3 \times 88}$

so $\qquad x_{OS} = 3 \dfrac{e^{0 \cdot 251 \times 88}}{e^{0 \cdot 3 \times 88}}$

$$= 3 \dfrac{e^{22 \cdot 088}}{e^{26 \cdot 4}}$$

$$= 3 \, e^{\, -4 \cdot 312}$$

As we started with three lesions the reduction in inoculum required is

$$\frac{3e^{-4\cdot312}}{3} = e^{-4\cdot312}$$

$$= \frac{1}{74}$$

This means that with this slightly increased rate of spread the amount an inoculum must be reduced is to one infected seed potato in every 74 square kilometres if the incidence of disease is not to increase!

These are very high rates of spread so let us consider a disease with one tenth the rate of spread, at 0·0251 per unit per day. Crop yield again stops at 75 per cent defoliation after 88 days. If a season occurs in which the rate of spread is increased to 0·030 per unit per day how much must sanitation be increased to hold losses at the same level? Following the same calculation, $x_O \, e^{rt}$;

therefore
$$\frac{e^{rt}}{e^{r't}} = \frac{x_{OS}}{x_O} = \text{Sanitation ratio}$$

$$\frac{e^{0\cdot0251 \times 88}}{e^{0\cdot03 \times 88}} = \frac{e^{2\cdot2088}}{e^{2\cdot04}}$$

$$= e^{-0\cdot4312} \simeq \frac{1}{2\cdot72}$$

i.e. inoculum has only to be reduced to about one third.

So far we have seen that sanitation is most effective as a control measure if r is low. This does not mean that it is only useful in diseases which naturally have a low rate of spread, but it is often most useful in conjunction with other control measures which reduce r.

The effect of increasing t will be similar to increasing r. This means that sanitation is most effective in diseases with either an extremely low rate of spread or those which have a very short period in which to spread.

Let us have a look at some of the bumps in the sigmoid growth curve. These are often due to such factors as seasonal sporulation or growth. If the crop being investigated is an annual it may only be affected by a single sporulation so it is important to know what happens during this 'bump'. (If the same crop is grown year after year the pattern may average out to the sigmoid growth curve we have so far considered.) A wide range of diseases show this periodic spore release. They include loose and stinking smut, ergot, all fruit-infecting diseases and many others.

Such diseases do not spread in a logarithmic manner, but their spread is linear i.e. analogous to simple interest not compound interest. In other words, the proportion of plants (dx) that will become infected during a short period of time (dt) where the infection rate is R can be expressed as

$$\frac{dx}{dt} = x_O R (1 - x)$$

where; x_O is the amount of original inoculum, and $(1 - x)$ is the proportion of plants left to infect.

(Although dx is thought of as a short period of time it could in fact be a whole growing season.) This means the amount of infection is proportional to the original amount of inoculum whatever the value of R.

If the rate of increase of a disease (dx/dt) is constant, $x_0 R$ must be constant, so that an increase in the rate R can be compensated by a reduction in inoculum x_0. For example if R is increased by $1/4$ how much will x_0 have to be reduced to stop the final level of disease becoming greater? The relative infection rate is now $5/4 R$. To keep $x_0 R$ constant the inoculum must be $4/5 x_0$, i.e. an increase in infection rate of $1/4$ can be compensated by a reduction of inoculum of $1/5$. Similarly an increase in inoculum x_0 of $1/4$ can be compensated by a decrease in rate of infection of $1/5$.

It is clear that a disease of this type is much more easily controlled by sanitation than a 'compound interest' type of disease. If however a series of sporulations are considered the 'compound interest' calculations again apply.

3.5 Latent period

This is the time taken from infection to the production of reproductive structures eg spores. The length of this time period will have a great effect on the rate of spread of the disease outbreak because the infection rate r will depend on the latent period, the number of reproductive propagules produced as a result of each infection and the percentage of propagules which infect another host.

The latent period is changed enormously by the environmental conditions and surprisingly little by other factors such as host resistance. An extreme example of the effect of environment is in the development of stripe rust in wheat. Under optimal conditions it takes about eleven days from infection to the production of spores but in the winter it can take as much as 120 days.

It is obvious that this is a major factor in determining whether a disease becomes epidemic or not. so that a knowledge of the effect of environment on the development of disease and a prediction of the coming weather conditions is of great value to a farmer in planning his crop protection regime.

3.6 In brief

Where growth of the pathogen is vegetative, or there is only one sporulation during the period under consideration so that the spread of the disease is linear, then sanitation can be a very effective control measure, but where there are many sporulations, i.e. spread is logarithmic, then sanitation can only be effective if the rate of spread of the disease r is low or the time during which it spreads is short. If the rate of spread is comparatively high, or the time during which the disease can spread is long, then the reduction of the rate of spread or time of spread is far more effective in controlling the disease than reducing the initial inoculum by sanitation.

As not infrequently happens agriculturalists have arrived at the same conclusions somewhat earlier than the scientists. Nurserymen for instance take great care to reduce inoculum by using selected disease-free seed which frequently has been treated with a 'dressing' to reduce further the likelihood of infection. Soil for seed boxes is usually sterilized as are the boxes themselves. These are the conditions in which pathogens are most likely to be in the stage of vegetative growth and the time in this environment is short. Both these factors make sanitation a practical and effective method of disease control. In the field most diseases will be showing logarithmic growth so that sanitation is less effective and methods of reducing the rate of spread r are used.

If r is reduced by one method or another then sanitation becomes practical even in the field, especially if it involves little effort and cost, as in the case of seed dressings. The rate of spread of the disease r can be reduced in a number of ways which include the use of fungicides and bacteriocides, growing resistant plants, as well as careful control of cultural practice.

The agriculturalist therefore attempts to balance the cost and benefit of these various control measures so as to produce the maximum profit on this crop and for future crops.

Chapter 4
Infection

4.1 Dormancy

Many plant pathogens produce resistant propagules to enable them to withstand inclement conditions. All that some propagules require to germinate is an adequate water supply whilst others require very specific conditions to initiate germination.

The period of dormancy that occurs prior to germination can be of two types: (i) Constitutive dormancy; in this the resting structure does not develop because of an innate property. This may be used to ensure the spore does not germinate at an unsuitable time of year. The simplest explanations of this sort of phenomenon is to imagine a necessary metabolic intermediary being slowly synthesized during the resting period, or an inhibitor being slowly broken down. (ii) Exogenous dormancy; in this situation germination is delayed because of unfavourable external conditions, either physical or chemical.

These two types are often linked and there is frequently more than one factor causing dormancy. For example, a spore may be dormant because of the presence of an inhibitor causing a metabolic block. This inhibitor can be leached away by free water in the external medium but leaching can only take place once the spore coat has become permeable. The spore coat becomes slowly more permeable throughout the winter, an effect perhaps accelerated by freezing. Germination will then occur in the following spring provided there is an abundant supply of water. Thus endogenous and exogenous dormancy mechanisms can be linked together to ensure germination at a suitable time of year and under suitable environmental conditions.

The time for which different types of resting structures will survive in a state of dormancy varies considerably and is very greatly affected by environmental conditions. It has been reported that paleozoic salt contained viable *Pseudomonas* spores and other workers have reported isolating bacteria from precambrian rock. There is considerable doubt about these findings because of the difficulty of not contaminating the samples with much more recent material in their extraction. More recent material investigated, faeces from 1500 BC and intestine from 1000 AD, were both found to be sterile. Even more recent material of accurately known age has been investigated and viable bacterial spores have clearly been shown to be present in material over one hundred years old. In soil from the roots of herbarium specimens of known age it was found that species of *Bacillus* are the most resistant and viable fungal spores were rarely found in material over fifty years old. From these studies it has been estimated that 90 per cent of the organisms in dry soil die every fifty years. At this rate commercial sterility would be achieved in about 1000 years. Under normally prevailing field conditions longevity of dormant spores would be only a very small fraction of this period.

4.2   Factors affecting survival

High temperatures are often used to sterilize soil, seed trays and other infected material but some fungi are surprisingly resistant to heat. For example, spores of *Colletotrichum* and *Verticillium* can survive for considerable periods at temperatures in excess of $50°C$ and the micro-sclerotia of *Verticillium albo-atrum* can be even more heat resistant. As soon as germination starts such organisms become much more sensitive to high temperatures and therefore the induction of germination prior to sterilization greatly increases its effectiveness.

The mechanism of heat resistance is not clear and a number of hypotheses have been proposed. It is known that the absence of water increases the resistance of many organisms to heat and it has been suggested that heat resistance is due to dehydration of the cytoplasm. This does not explain all the observations but there does seem to be a fairly general relationship between water content of spores and their longevity, if not always with their heat resistance. For instance, spores with a water content of more than 20 per cent usually have a dormant life span measured in days whereas those with a water content less than 20 per cent usually have a dormant life span measured in years. There also seems to be some relationship between heat resistance and mineral content. High levels of divalent cations, especially calcium, in the growth medium seem to enhance the heat resistance of the resulting spores and limiting calcium in the medium tends to reduce their heat resistance but longevity does not seem to be affected by mineral ions in this way. No one has yet put forward a tenable hypothesis to explain these phenomena. It has been noted that there is also a relationship between the dipicolinic acid content of many spores and their heat resistance, which between certain limits is nearly linear. Heat-resistant enzymes have been extracted from bacterial and fungal spores but the difference in structure between them and their heat-labile counterparts has not yet been demonstrated.

Spores of some bacteria and fungi are exceedingly resistant to low temperatures. *Bacillus subtilis, B. mesentericus, Mucor mucedo, Rhizopus niger, Aspergillus niger* and *Penicillium glaucum* all have spores which survived experimental temperatures of between $0·0075°$ K and $0·047°$ K at 10^{-6} mm of mercury. *Mucor mucedo, M. racemosus, Aspergillus niger* and *A. glaucus* all had spores which germinated two years after being treated with liquid air $(-190°C)$ for 492 hours followed by liquid hydrogen $(-253°C)$ for 77 hours. Many plant pathogenic micro-organisms are killed by such intense cold and several theories have been put forward to explain the cause of cold injury. One theory suggests that the damage is due to the build up of electrolytes as the result of relatively pure ice crystals forming in the cell. Another theory suggests that damage is due to rupture of membranes by ice crystals. It has been found that extremely low temperatures can cause up to 90 per cent of the intracellular water to freeze, i.e. if the ice crystals are of pure water the solutes in the cell will be ten times more concentrated. It is not merely temperature and the time spent at that temperature that is important, the rate of cooling and then

heating can be critical. Lowering the temperature slowly generally tends to reduce
the mortality, possibly because intracellular water is excreted and polysaccharides
and proteins are hydrolysed fast enough to keep the cell contents from freezing. If
freezing of the cell contents did occur, slow cooling would produce larger ice
crystals than rapid cooling. This may not be important if even rapid cooling
produces ice crystals large enough to damage cell membranes irreparably. In a few
instances fatalities are increased by slow cooling so perhaps the cause of damage
varies in different circumstances. The medium in which the cooling takes place will
also affect survival. High concentrations of mineral salts tend to increase mortality
whilst other materials in the medium such as glycerol, can aid survival.

A study of the rate of cooling, surface area to volume ratio, permeability and
Q_{10} (the effect of each ten degree centigrade change in temperature) have led to a
prediction that if the water content of a spore is reduced to below 10 per cent then
no intracellular ice will form. Therefore the relatively low water content of many
spores could account for their tremendous tolerance of low temperatures. At these
very low temperatures (below $-200°C$) the following effects would occur: (i) cell
constituents solidify (with or without the formation of ice crystals); (ii) most gases
including N_2, O_2, CO_2 are liquified or solidified; (iii) dissociation and ionization of
molecules is suppressed; (iv) chemical reactions would proceed at about one
millionth the rate at which they take place at $20°C$; (v) the molecular state of some
cell components, eg cytochrome, would be changed. Despite these drastic changes
many spores survive and in some cases survival is greater at these low temperatures
than at more normal temperatures. It would therefore seem likely that many of
these effects are in fact compensating for each other.

Moisture is most important in the survival of spores and resting structures, and it
greatly modifies the action of other environmental factors. It is possible that some
extreme environmental conditions, such as high temperature and solutions of high
osmotic pressure kill spores by desiccation.

Paradoxically many heat resistant spores are more tolerant of high temperatures
when the water content is low. It is for this reason that steam sterilization is
generally more effective than dry sterilization. In some cases even vegetative cells
have their survival potential increased by drying and this is utilized by freeze-
drying material to be stored in culture collections. This is not always the case and
some species far from having their survival enhanced by drying are killed by the
drying itself. It seems that the speed and conditions of drying are important in
determining the survival of the organism. There is however a spectrum, with some
spores surviving best under dry conditions whilst others require fairly humid
conditions for optimal survival. For example, uredospores of *Phymatotrichum
coronata*, conidia of *Helminthosporium oryzae* and many sclerotial resting
structures survive best at humidities of 20-30 per cent and *Puccinia graminis* survives
best at about 40 per cent humidity, whilst spores of *Erysiphe graminis, Erysiphe
polygoni, Oidium heveae* and *Tilletia tritici* have maximum survival at humidities
in excess of 70 per cent.

Radiation is deleterious to the survival of spores but once again there is
enormous variation in their sensitivities. With the longer wavelengths of ultra-violet
and visible light the variation in sensitivity is clearly related to the degree of

pigmentation, the pigments absorbing the radiation so that it does not penetrate to the protoplasm. The heavily pigmented spores of *Aspergillus niger* for example can withstand 10-20 times the dose of ultra-violet radiation that can be withstood by the unpigmented spores of *Neurospora*. When pigmented spores germinate they will naturally lose their resistance to ultra-violet and visible light unless the germ tube itself is pigmented, something which is comparatively rare. With shorter wavelengths such as X-rays pigmentation has of course no effect and here there is variation between the resistance of spores from different species quite unrelated to their pigmentation. Resistance to X-rays is perhaps related to the stability of vital cellular macro-molecules, due perhaps to disulphide bonds in protein molecules; certainly many radiation resistant spores contain large amounts of the sulphur-containing amino acids cysteine and cystine.

4.3 Germination

In many cases all that is required to stimulate the germination of a spore or propagule are conditions suitable for vegetative growth. It is frequently possible to stimulate germination with conditions that will not sustain continued vegetative growth. For example many non-resistant spores can be stimulated to germinate simply by putting them in distilled water. In other instances a stimulant that is not necessary for vegetative growth is required to initiate germination.

Although it is common for many bacteria and coprophilous fungi to require heat treatment before they will germinate, such as they would receive in the gut of a mammal or bird, it is surprising that some plant pathogens also show increased germination after heat treatment. *Ustilago striiformis* shows maximum germination after the spores have been treated for about thirty days at 32-35°C and *Neurospora* spores are stimulated by even higher temperatures. Cold treatment is less frequently required by micro-organisms to trigger germination than by seeds of flowering plants. Nearly all the micro-organisms that do show cold stimulation are plant pathogens and it is almost certainly a mechanism to synchronize growth of the pathogen with that of the host.

There is a wide variety of chemicals that have been found to stimulate the *in vitro* germination of spores. Some of these are required for only a short period to initiate germination and can be removed prior to germination without stopping growth, whilst other compounds have been found to be necessary right up to the time of germination and sometimes they are required after this, during early vegetative growth. In these latter circumstances they tend to be regarded as nutrients rather than germination stimulants.

Wetting agents frequently stimulate spore germination probably by increasing the availability of water to the protoplasts of the many spores which have a waxy coat to resist desiccation. It is not only detergents that have a wetting action. Compounds such as ethanol and acetone also reduce surface tension but it is not certain if it is their wetting action or their action as fat solvents which stimulates germination because other organic solvents, such as chloroform and ether, can also increase the percentage germination of some spores. As the latter have little effect on the surface tension of water it is likely that it is their action as fat solvents that is important, but this does not exclude the importance of water. It is possible that the solvents increase the permeability of fat rich spore coats so that once again water is more available to the protoplast. Spores stimulated to germinate by organic solvents include the uredospores of *Puccinia*, the ascospores of *Neurospora* and the conidia of *Aspergillus* as well as the spores of several smut fungi. Wetting agents are also important in the early stages of infection by overcoming the tendency of water to remain as discreet droplets on the waxy cuticle of a leaf. Such a droplet is likely to run off carrying the spores with it and even if lodged the droplet presents difficulties to infection. Those spores requiring a solid substrate will not germinate and the small surface of contact will limit the diffusion of chemical stimulants from the leaf surface into the drop. Even if germination is successful the hyphae will then have difficulty penetrating the air–water interface and will in all probability have to cross an air gap before they reach the leaf surface.

Many of the organic compounds produced by plants stimulate spore germination. A few of the earliest shown to have this action include ethyl esters of acetate, malate, and citrate. Occasionally spore germination stimulants have been found to be involved in the specificity of a pathogen for a particular host. For example benzaldehyde is a constituent of almonds and has a stimulatory effect similar to the crude host extract, on a number of almond pathogens.

Host plant exudates can not only trigger spore germination but in some instances attract the pathogen to the host. Ethanol has been shown to attract zoospores of *Phytophthora cinnamomi* and as this pathogen attacks trees in waterlogged conditions it is likely that the ethanol produced by anaerobic respiration of the roots is attracting the zoospores which then successfully infect them.

It is not only organic compounds that are important in stimulating spore germination. Hydrogen ion concentration is most important in breaking dormancy. There are a number of instances where the mechanism by which a compound stimulated germination was studied in great detail only to discover the effect was due to a change in the hydrogen concentration and not to any special chemical property of the compound. A great many fungal spores are stimulated to germinate by the presence of other inorganic ions. For instance *Ustilago* has a requirement for sodium carbonate and *Puccinia glumarum tritici* is stimulated by NH_4^+ and smuts are stimulated to germinate by calcium nitrate. Many spores show increased germination in the presence of NO_3^-, PO_4^{3-} and SO_4^{2-}. Penetration as well as germination can be affected by inorganic ions. *Puccinia coronata*, for example, requires Zn^{2+} before it will form an appressorium (the structure produced to penetrate the cuticle and epidermis) and this it obtains from the surface of the host leaf. It seems most unlikely that the effect of these ions is due solely to the alteration of pH that they undoubtedly bring about.

It is not only micro-organisms that are affected by the material liberated from plants. The seeds of many flowering plant parasites such as dodder (*Cuscuta*), broomrape (*Orobanche*), and witchweed (*Striga*) germinate only in the presence of a suitable host. Their germination is triggered by specific substances liberated by the host roots, thus ensuring germination only occurs when a suitable host is available. Even after broomrape seed has been stimulated to germinate with host root extract it is still not able to attack non-host roots when placed in contact with

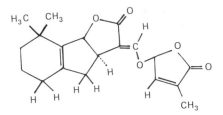

Figure 4.1 Strigol

them, so germination is not the only specificity in this host—parasite relationship.

Recently the structure of the compound liberated by cotton roots which stimulates *Striga* to germinate has been elucidated. This compound called 'strigol' is active in concentrations as low as 10^{-11} Molar.

Although many plant materials stimulate germination of spores there are some that are very active in inhibiting it. The oils from orange peel, onions and garlic are all strongly inhibitory not only to germination but to subsequent mycelial growth of many fungi. The waxy cuticles of apples and many plant leaves contain both water-soluble and fat-soluble germination inhibitors. In some instances this may form the basis of disease resistance. For instance, certain varieties of gram (*Cicer*) appear to be resistant to *Mycosphaerella rabiei* because of a secretion of malic acid which *in vitro* reduces spore germination and inhibits hyphal growth. Similarly protocatechuic acid produced in the scales of some onions seems to confer resistance to *Colletotrichum circinans* by inhibiting spore germination. Substances inhibitory to the growth of a number of fungi can be washed from the surface of oak leaves with both water and ether. Interestingly these substances do not inhibit the growth of oak mildew caused by *Microsphaera alphitoides* which appears to be resistant to inhibitors liberated by its natural host.

Plant roots release into the soil comparatively large amounts of organic material including sugars, amino acids, proteins, organic acids and many others. The region around the roots of a plant (rhizosphere) soon becomes colonized by large numbers of micro-organisms using these substances to supply their energy requirements. The competition for these nutrients is intense and the organisms able successfully to use them are by and large very specialized rapid growing saprophytes. To aid their competitive ability many of these saprophytes produce compounds inhibiting the growth of other micro-organisms. It is these compounds that are utilized by man as antibiotics. Parasites adapted for growth within a host plant are usually poor competitors in the saprophytic environment surrounding a plant root so that they will have little vigour with which to attack the the host plant. The rhizosphere microflora therefore form the outermost defence of a plant root against invasion by a pathogen. The same sort of phenomenon occurs on plant leaves and stems but to a much more limited extent because it can take place only on the leaf and stem surface, and the surrounding area cannot be affected in the same way as soil around plant roots.

4.4 Predisposition

As we have mentioned environmental conditions can alter the susceptibility of a
plant to infection. However it is sometimes difficult to distinguish between conditions favouring the growth of the pathogen and those increasing the predisposition
of the host to infection, eg the formation of dew on a plant increases the likelihood
of infection taking place. Is this because it favours the growth of the pathogen by
providing free water for germination of spores or because of a predisposition of the
host to infection brought about by the cooling of the tissue during dew formation?
(It is well established that the cooling of host tissue frequently increases its
susceptibility to infection.) It is considered that both these phenomena have an
effect but the presence of free water for germination is probably the more
important.

Micro-climate is an example of a factor affecting the germination of spores and
the survival of the pathogen after germination and prior to infection. It also has a
less direct effect on the pathogen by altering the metabolism of the host and thereby causing an increase or decrease in its resistance to invasion. A host may be
attacked by a pathogen that can enter only through stomata and it will therefore be
more susceptible when the stomata are open than when they are closed. So
conditions favouring transpiration will also favour infection. There are many factors
other than micro-climate that will influence infection through an alteration in the
host's metabolism, many in a subtle and not at all obvious way.

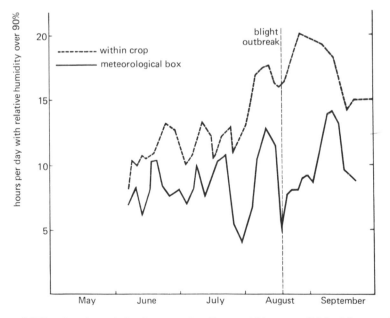

Figure 4.2 Showing the variation between the climate within a crop (Majestic's potato) and
outside the crop, the outbreak of *Phytophthora* blight being clearly related to the climate
within the crop.
(after Hirst, 1958)

The term micro-climate rather than climate has been used because the conditions occurring on a small scale around the crop may not be those occurring a few feet away in a meteorological recording box. A study was carried out on the conditions within a potato crop and on the conditions four feet above the crop in a normal meteorological recording box and how these conditions related to the incidence of blight in the crop. It was found that there was a good correlation between the conditions within the crop and the outbreak of blight but little or no correlation between the conditions outside the crop and the outbreak of blight. Blight followed a very marked increase in the number of hours per day that the humidity within the crop was more than 90 per cent. The hours of high humidity outside the crop had at this time risen very little, i.e. the crop itself was producing a micro-climate conducive to a blight outbreak. It has been noticed that the rise in humidity within the crop coincides with, and is almost certainly due to, the potato leaves forming a closed canopy over the ground. The crop itself was therefore producing conditions favouring infection either by increasing the survival of the *Phytophthora* spores or by increasing the predisposition of the potato plants to infection, or both.

Fortunately the conditions so ideal for the development of blight will hinder the spread of the disease. The closed canopy limits wind dispersal and in any case the spores are not released in large numbers under humid conditions, but remain attached to the sporangiophores. This means the disease will develop rapidly under a closed canopy but will spread relatively little outside it. The disease itself however results in the break up of this canopy as the leaves yellow and die. This causes a fall in the humidity and the release of spores in extremely large numbers to be spread by the wind.

Not surprisingly plants, like animals, vary with age in their susceptibility to infection. There are many examples of plants increasing their resistance to infection as they get older. Many seedlings are attacked by species of *Pythium*, particularly under moist growing conditions, causing the complete collapse of the stem tissue just above ground level and resulting in the rapid death of the seedling (damping off) but *Pythium* is rarely successful in attacking older plants even under the most suitable environmental conditions. There are many other examples of increasing resistance with age; barberry to the basidial stage of *Puccinia*, wheat to the uredial stage of *Puccinia*, rye to *Fusarium*, lettuce to *Erwinia*, tobacco to *Peronospora*, and potato to *Phytophthora* to mention just a few.

At the other end of the scale there are examples of increasing susceptibility with age: peach and apricot fruits to *Sclerotinia,* cucumber to *Pseudoperonospora* and lettuce to *Erysiphe*. There are however some examples of resistance to a pathogen by young plants which become susceptible in middle life, and resistant again in old age: for instance, potato tubers become susceptible to *Erwinia* in middle life as do beans to *Uromyces.*

As plants are continually growing and producing new young tissue most of these results refer to the age of the tissue rather than the age of the host plant. Obviously these changes in susceptibility are the result of changes in the metabolism of the host cells as they age.

There is also variation in susceptibility depending on the season or the time of day. Plum trees for example are more resistant to *Stereum* during the summer months than during the rest of the year and chestnut is more resistant to *Nectria* in the spring and summer than in the autumn and winter. Spruce and silver fir are more resistant to rot in the autumn and early winter than in the late winter and spring. A number of reasons have been suggested for these variations ranging from different water content to changes in hormone levels in the host but there is as yet little evidence to back any of the theories.

Diurnal variation in susceptibility (or resistance) is even more common than seasonal. There are so many diurnal effects in plants that it is difficult to know which are related to disease resistance and which are not. Fluctuations during the day include the following: leaf movements, stomatal opening, nuclear division, cell elongation, as well as changes in wetability of leaves, occurrence of plasmodesmata (cytoplasmic connections between cells), guttation (secretion of water from the tips of leaves), organic acid concentration, carbohydrate concentration, protein content, hormone balance, mineral and water content. As we have seen plant pathogens also show diurnal variations in the formation of spores, discharge of spores, germination of spores, and the formation of appressoria. These diurnal variations by the pathogen could be in direct response to the environment or they may be in response to alterations in the host brought about by the environment. This, of course, is predisposition.

In some instances the relationship is clear. The diurnal opening of the stomata of both soybeans and peaches has been shown to be related to the susceptibility of these plants to infection, being greatest when the stomata are open and at a minimum when they are closed. In contrast to this *Puccinia triticina* stimulates the closure of host stomata prior to forceful penetration. It would be tidy to consider the closure of the stomata an unsuccessful attempt by the host to stop penetration but there is little evidence in favour of this hypothesis. High or low temperatures frequently predispose a plant to infection and the following are a few of the large number of examples of heat increasing the susceptibility of plant tissue to invasion by a pathogen: apple to *Botrytis*, bean to *Erysiphe polygoni* and *Uromyces*, potato to *Erwinia* and *Fusarium*, sugar cane to *Colletotrichum falcatum* and wheat to *Puccinia*. Occasionally heat may decrease susceptibility of a plant to infection. Tomato roots dipped in hot water become more resistant to subsequent infection by *Fusarium lycopersici*, and bean leaves heated to 55°C for ten minutes are more resistant to infection by tobacco mozaic virus. Cold may also occasionally reduce the susceptibility to subsequent infection eg wheat and rye are more resistant to *Tilletia tritici* and *Ustilago nuda* if previously exposed to low temperatures (vernalization). Usually however, low temperatures tend to increase susceptibility to infection.

The effect of night temperatures on the susceptibility of rice to blast caused by *Piricularia* has recently been investigated by workers in Madras. A large number of strains of rice were taken, some considered to be resistant to blast and some considered susceptible. They were grown with a night temperature falling to just over 20°C and the various strains exhibited the expected resistance or susceptibility to the pathogen. When they were grown with the night temperature always above 30°C all strains both 'resistant' and 'susceptible' were found to be resistant to the pathogen, but if the night temperature fell to below 15°C even the 'resistant' strains started to become infected. Meteorological forecasting is now used to warn farmers of expected cold nights to enable them to spray their crops with fungicides as a protective measure.

The change in susceptibility of rice to *Piricularia* would seem to be related to the nitrogen metabolism. At the higher night temperatures the amount of free nitrate in the tissue is lower than when the plants are exposed to low night temperatures, probably due to the increased nitrate reductase activity detected under these conditions. Providing high levels of nitrate in the form of fertilizers increases the susceptibility of rice to blast especially at low night temperatures.

Humidity would seem to have relatively little effect on the predisposition of the host to infection. The effect of humidity is largely on the pathogen directly but water availability in the soil and consequently the water content of the host does have a marked effect on susceptibility. Almost without exception the higher the water content of the host the more susceptible it is to infection: for example, apple twigs and potato are more susceptible to *Erwinia,* and peas and cucumbers are more susceptible to *Pseudomonas,* when the water content of their tissues is high. Similarly, tobacco is more susceptible to bushy stunt virus and potato is more easily colonized by *Botrytis, Oospora* and *Phytophthora* under wet conditions.

There are a few exceptions to this generalization, which include *Pythium* on potato tubers, *Peronospora* on beet, tobacco mozaic and tobacco necrosis viruses on beans. These diseases are said to be favoured by dry conditions prior to infection.

Light can vary both in intensity and in duration (day length) and both variations have an effect on the predisposition of plants to infection. However, these effects are not constant, depending on species of host, species of pathogen and on the other environmental conditions.

Reduced light intensity prior to inoculation increased the susceptibility of lettuce and tomato to *Botrytis,* of tomato to *Fusarium,* of elms to *Ceratocystis* and of tobacco to spotted wilt virus, possibly by reducing the nutritional status of the plants. On the other hand the exposure to sunlight for a few hours of seed potatoes cut ready for planting increased subsequent decay. This could be due to the inhibition of wound healing by ultra-violet light. Certainly gamma radiation has been found to inhibit the suberization that normally occurs after a potato tuber is cut and which helps to protect the tuber from invasion by pathogenic organisms and gamma radiation has also been found to increase the susceptibility of resistant wheat and oats to *Puccinia.* It has been shown that ultra-violet radiation can in some instances increase the susceptibility of the host to infection, eg broad bean attacked by *Botrytis.* The effect of ionizing radiation on host resistance is not just due to gross damage to the host tissue because ultra-violet treatment had no effect on the susceptibility of broad bean to *Uromyces.* X-rays have also found to alter the susceptibility of hosts to infection, but again no generalizations can be made about increased or decreased susceptibility. For instance, the susceptibility of tomato to *Fusarium* is increased or decreased by X-rays depending on the dose.

The effects of day length are also variable. Short days favoured the infection of tomato by *Fusarium* but intermediate day lengths favoured the infection of currant by *Cronartium.* That day length caused different responses in different host plants is not surprising because many of the developmental processes (and therefore the metabolic processes producing them) are sensitive to day length, eg flowering and these are likely to affect the susceptibility of the plant to infection.

Mineral nutrition is one of the most important environmental factors affecting the predisposition of a plant to infection. A tremendous number of experiments have been carried out on the effects of nutrients on infection, far too many to even start listing specific examples. The general findings have been that high levels of nitrogen increase the susceptibility of a wide range of hosts to fungal, bacterial and viral infections. There are naturally exceptions and there are a few instances of high nitrogen levels decreasing susceptibility. Phosphorus is thought to increase the susceptibility of a fairly wide range of hosts to viral infection but it also increases the resistance of some plants to fungal infection. High potassium has also been found in a number of instances to reduce the susceptibility of plants to fungal infection.

Other minerals usually have less effect than these three and the findings vary more from host to host and pathogen to pathogen. Silicon has in some instances increased the resistance of cereals to fungal infection possibly by increased mechanical strength of the cell wall thus offering greater resistance to penetration. Boron (in sub-toxic doses, of course) has been found to increase the resistance of a number of hosts to fungal attack, as have manganese, copper, zinc and lithium in suitable doses.

It is likely that the ratio of nutrients is even more important than the absolute concentration of any one element. There are however examples of changes in susceptibility where the ratio of nutrients is unaltered but the total concentration of nutrients is changed. Very large changes in nutrient level are required to produce any alteration in susceptibility and many people consider the osmotic effects of these massive changes in mineral ion concentration to be more important than the nutrient effects. For example, massive increases in the nutrient concentration supplied to beans grown in water culture increased their susceptibility to *Erysiphe*. Hydrogen ion concentration has a considerable effect on the infection process, but this is not the same for all host parasite combinations. Sometimes the increased infection takes place at low pH, sometimes at high pH and sometimes maximum infection occurs at around neutrality. It is difficult to determine whether the effect of pH is on the host (predisposition) or on the pathogen. Pre-treatment of the host at a given pH, before returning it to a standard pH at the time of inoculation, might be one way of separating these two phenomena, but rapid changes of pH prove to be difficult with large masses of tissue and allowing equilibration at the standard pH would tend to allow the metabolism to change from that at the experimental pH. Changing the pH could also induce effects that were the result not of the final experimental pH itself but caused rather by the fact of changing the hydrogen ion concentration.

It seems highly likely that hydrogen ion concentration is very important in the predisposition of a plant to infection but because of the experimental difficulties (such as those already mentioned and the nutrient or mineral value of the buffer used to control the pH) there are few conclusive experimental findings.

Wounding can in many instances facilitate the entry of the pathogen and it can therefore be considered as predisposing the host to infection. Wounding can be of many sorts, including cutting, crushing, heat damage, frost damage, chemical damage, probing by sap suckers, etc. Damage causing the death of a region of tissue can increase susceptibility by providing a suitable dead organic substrate for the pathogen to grow on prior to attacking the healthy tissue, i.e. increasing the inoculum potential. Wounding also alters the metabolism of the host and thereby alters the susceptibility of tissue some distance from the wound. These alterations are likely to be on any one or more of the following: the hormone balance, the osmotic state of the tissue, the carbohydrate balance, the water balance, protein synthesis, respiration, photosynthesis, translocation and transpiration, all of which are likely to alter the ability of the tissue to withstand invasion by a pathogen.

Wounding does not always result in increased susceptibility. The removal of fruit or the removal of growing points and young leaves increases the resistance of tomato to attack by *Macrosporium,* and the removal of flower buds from cotton tends to increase its resistance to a number of pathogens. This might be a hormonal effect (most of the tissues mentioned are the sources of a number of plant hormones) or perhaps the removal of carbohydrate sinks increases the level of carbohydrate in the rest of the plant which increases resistance. Certainly complete defoliation and the subsequent fall in carbohydrate concentration frequently increases the susceptibility of a plant to infection. This appears to support the latter hypothesis but the results of complete defoliation are so traumatic that many other factors are likely to be operating, and such results should be considered in this light and not be thought to be conclusive. In contrast, the removal of mature leaves from plants infected with *Verticillium albo-atrum* tends to increase their resistance to infection, an effect perhaps due to the fact that *Verticillium* causes a wilt disease in which the host shows signs of water stress stemming from either a limited water supply or excessive water loss. Defoliation would reduce water loss and give the host a better opportunity to produce defence reactions. Again, such a conclusion is not clear cut. The removal of leaves could cause the operation of a more subtle mechanism which limits the growth up through the stem of this root-infecting fungus.

With grafting it is generally considered that the two parts of the graft retain their original resistance after the graft is established (i.e. ignoring the effect of wounding in producing the graft). There are as always a few exceptions. In those instances in which the root stock reduces the vigour of the shoot, or where the shoot reduces the vigour of the root stock, the resulting plant is not surprisingly more susceptible to infection than either of the plants comprising the complete host. A far more specific interaction takes place in some grafts. Grafting sweet cherry on to a mahaleb root stock results in the shoot being completely resistant to the buckskin virus. The inclusion of just a single sweet cherry root causes the entire shoot to become susceptible. This shows that it is not the mahaleb roots producing a compound inhibitory to the virus but the sweet cherry roots producing one or more compounds which when translocated up into the shoot cause the shoot to become susceptible to virus infection.

Infection by one parasite frequently increases the likelihood of infection by another. This can result from a number of causes, the two most important being the production of suitable infection sites and the alterations to the host metabolism which may predispose the plant to further infection.

Prior infection however often increases the resistance of a plant to subsequent infection. The first infection stimulates the production of fungitoxic compounds which inhibit the growth of any subsequently invading pathogen. Such compounds are collectively called phytoalexins. Increased resistance to subsequent infection can also be due to the laying down of barricade tissue within the host which forms a barrier to both the initial pathogen and to any following attack. Increased resistance is in some instances due to the production of fungitoxic substances by the pathogen itself, inhibiting the growth of any other micro-organism attempting to infect the host plant.

The reverse of this process can also take place, where the initial pathogen liberates substances which diffuse through the host tissue damaging the cells and lowering their resistance to subsequent infection. In some cases the pathogen liberates substances so toxic to the host cell that they are killed in advance of the growth of the invading hyphae. An example of this is the case of soft rots caused by organisms such as *Sclerotinia fructicola,* where the hyphae actually grow saprophytically on dead organic material previously killed by fungal exudates. This, not surprisingly, leaves the host plant liable not only to colonization by specialized saprophytes, which will do little harm being limited to the tissue already dead, but to facultative parasites which can colonize the dead tissue and then grow out into the surrounding living tissue.

Chemical predisposition can be caused by chemicals applied to the host either deliberately or accidentally which are not used by the host as nutrients and which affect the susceptibility of the host plant to infection.

Copper (applied for instance as Bordeaux Mixture) increases the susceptibility of potato to *Penicillium, Fusarium* and to a number of bacterial infections. It increases the susceptibility of oranges to *Diplodia* and of celery to *Anatospora.* In these examples the predisposition of the host to infection seems to overcome the fungitoxic effects of the copper. Pesticides (insecticides, moluscicides, herbicides etc) can also increase the predisposition of a plant to infection. The insecticide DDT and the herbicide 2,4-D both increase the susceptibility of tomato to *Fusarium* wilt; 2,4-D also increases the susceptibility of wheat to *Claviceps* and other related herbicides can increase susceptibility of some plants to infection, eg 2,4,5-T increases the susceptibility of tomato to infection by *Didymella.* Many metabolically active substances can predispose a plant to infection. The respiratory inhibitors 2,4-dinitrophenol, thiourea, and sodium fluoride cause resistant tomatoes to become susceptible to *Fusarium.*

Other classes of compound also affect the predisposition of plants to infection. The emulsifying agents used in a number of sprays can increase the susceptibility of some plants to infection, eg broccoli to infection by *Alternaria.* A great many compounds not deliberately applied to plants can affect their predisposition to infection. These include such things as smog, and air-borne and water-borne industrial wastes, but the difficulty of demonstrating the direct rôle of such compounds in an agricultural situation often makes it impossible to control their discharge by legislation.

In addition to the ten different factors predisposing plants to infection which have been mentioned there are still others of somewhat less importance that have not been discussed here. The whole range of relevant factors interact one with another to produce compound or synergistic effects. Singly or together they must alter the metabolism of the host making it more suitable for the growth of the pathogen. This might be achieved by making the environment more suitable initially, thereby making infection easier, or by inhibiting the host from producing a defence reaction which would limit the later growth of the pathogen. For example, the copper treatment of the cut surfaces of seed potatoes that increases susceptibility to infection probably acts by slowing the suberization process that is a normal part of wound healing. This inhibition of wound healing is itself an effect rather than a cause and there must be an alteration in the metabolism of the cells near the wound causing them to have a slower than normal healing response.

4.5 Penetration structures

Factors affecting penetration have been discussed but no mention has so far been made of how the pathogens attempt to enter the host. Whether or not a fungal pathogen enters a plant through a wound, a stoma, some other opening in the epidermis such as a lenticel, or whether it penetrates straight through the epidermis, hyphae must pass through cell walls if they are to penetrate host cells. There are two ways in which such penetration may be achieved, either by mechanical penetration or by the chemical solution of part of the cell wall. Although many plant pathogens produce cellulolytic and pectolytic enzymes it is not considered that they play a very important part in primary penetration. Since the end of the nineteenth century there has been a steadily growing body of data demonstrating the mechanical penetration of host cell walls by pathogenic fungi.

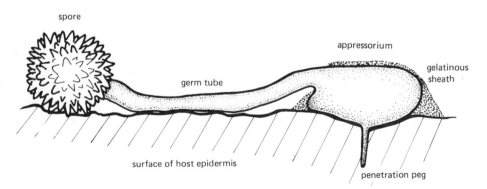

Figure 4.3 The penetration structure produced by some fungi to enable them to penetrate the surface of the host.

Amongst the early work was a detailed description of penetration of *Colletotrichum* into bean pods and this has since been confirmed and found similar to a large number of pathogens penetrating a wide range of tissue. It was observed that the spore germinated on the surface of the pod and the germ tube grew out across the tissue closely conforming to the surface. The germ tube became attached to the surface by a gelatinous sheath and the spore itself seemed to become attached in some way although no gelatinous covering was visible over the spore. The tip of the germ tube then turned in towards the surface of the host and continued growth caused the older part of hyphal filament to arch upwards away from the surface of the bean pod. This growth of the hypha towards the surface of the host caused a small identation in the wall of the cell under the hyphal tip. A small protruberance was then produced at the tip of the hypha (now called an appressorium) which dented the host cell wall even further. Finally this protruberance developed into a fine peg which punctured the cell wall. After penetration the hypha rapidly thickened and became branched but the region passing through the cell wall remained very fine, hardly enlarging its diameter at all.

There followed many other similar accounts, for example the penetration of *Botrytis cinerea* into *Vicia faba* and the infection of barberry by *Puccinia graminis*. Some fungi, such as the smuts, have a simpler mechanism of penetration than *Colletotrichum*. *Ustilago maydis* is the smut that has been most extensively investigated; it just produces a very fine tip to the hypha which punches through the cuticle and cell wall, no appressorium of any sort being formed. Penetration in this genus is largely limited to the very young cell of the host and this possibly reflects the simplicity of the penetration structure.

It is still not certain whether enzymic degradation of the cell wall ever plays a part or whether penetration is always purely mechanical. This is a difficult field of study. For example *Verticillium albo-atrum* produces cellulases and pectinases

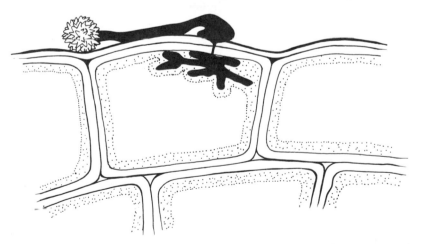

Figure 4.4 Penetration of a host cell by a pathogenic fungus and the formation of haustoria through which the fungus absorbs nutrients. The haustoria push into the cytoplasm without initially causing the rupture of host cell membranes.

in artificial culture when cellulose or pectin are the only available respirable substrates but the presence of other substrates, such as glucose or sucrose, inhibits the production of these enzymes. During the infection of the living host (eg lucerne) little if any cellulase is produced but after the death of the host and the exhaustion of soluble carbohydrates, cellulase is produced and the host cell walls are used as a respirable substrate. This is clearly consistent with the effects in culture of soluble carbohydrates inhibiting the synthesis of cellulolytic enzymes. Are there then enough soluble carbohydrates on the surface of the host cell and within the host cell walls to inhibit the production of cellulase? If there are not, are cellulases produced in undetectably tiny amounts just at the tip of the penetration peg? As yet we have no definite answers to these questions.

Penetration frequently occurs in plants that are resistant to that particular pathogen, growth of the pathogen being stopped after penetration. This suggests that resistance to penetration is rarely the sole mechanism by which plants resist attack by potential pathogens.

4.6   Penetration sites

Natural openings and regions where there is no cuticular or suberized covering to the surface cells occur in a number of regions of a plant. These include stomata, hydathodes, lenticels, root hairs, and the point of exit of lateral roots. The entry of pathogens through these openings varies. It has been observed that fungal or bacterial spores landing on a drop of dew or exuded sap at the entrance to a stoma will be drawn into the stoma as the droplet evaporates. In other instances a fungal spore may germinate on the surface of a plant and grow over the surface until a suitable opening is found. The hypha passes through this opening and growth is frequently continued for some time within the air spaces of the host tissue before a cell is penetrated. In some instances a specialized structure is produced at the opening, as in the case of *Puccinia graminis* attacking wheat. The spore germinates and grows over the surface of the wheat leaf until it reaches a stoma, when an appressorium is produced that fits exactly over the guard cells and a penetration tube pushes its way between the guard cells. Once in the stomatal chamber the hypha swells up into a sizeable vesicle and where this touches the surrounding mesophyll cells fine peg-like penetration hyphae are produced which push through the cell walls. After penetrating the mesophyll cells these hyphal pegs enlarge and branch into knobbly or finger-like haustoria. These are the absorptive region of the fungus drawing nourishment from the host cell which survives for a considerable time after infection.

Some fungi, such as *Phytophthora infestans* attacking a potato plant, can enter either through the stomata, or directly through the cuticle and epidermis, but many pathogens can enter a host only through a natural opening, eg *Cercospora beticola* attacking beet. In fact some pathogens are unable to penetrate a stoma if the guard cells are closed and as previously mentioned this leads to great variations in the susceptibility of the host under different conditions. An example of such a disease is the stem rust of wheat which can only achieve infection whilst the wheat stomata are open during transpiration.

Most of the pathogens entering a host through stomata or lenticels also enter through wound sites, eg *Streptomyces scabies* – the common scab of potatoes, *Sclerotinia fructicola* – the brown rot of stone fruits, *Penicillium expansum* – the blue mould of apples, *Erwinia caratovora* -- the bacterial soft rot of potato and other vegetables.

Normally when a fungus grows through the loose cells of a lenticel or wound-healing tissue, the tissue is stimulated to produce tightly packed elongated cells which frequently develop into cork and may or may not be successful in stopping the spread of the fungus. (This will be considered in more detail when dealing with host reaction.)

Wounds, of course, produce an ideal site for infection by many pathogens, the loose callus tissue composed of thin-walled parenchyma cells is easily penetrated. A wound before this stage of healing has occurred is an even more favourable site for the establishment of an infection. With nutrients still leaking out of damaged cells, and from the ends of xylem and phloem if any vascular strands have been damaged, and with the humidity maintained high by the loss of water from the host tissues it is not surprising that a large number of infections occur through wounds. Many of these wounds are produced by cultivation (pruning, lopping, harvest, etc), whilst others are produced by animal predators (including herbivores as well as insect and nematode pests) and by fire. It may come as a surprise that fire is probably the most important of these in the production of infection sites in forest trees. The practice of burning off under-brush is thought to result in untold damage by later infection of the fire wounds.

Whether one considers self-inflicted damage (such as the splitting of bark due to growth, the breaking of root hairs as the root elongates and the damage to the cortex of roots by the growth of laterals) as wounding is a question of semantics, not plant pathology. From the point of view of the pathogen it is as suitable for colonization as wounds from any other cause and the comments made about other wounds are entirely applicable to self-inflicted ones.

Penetration by pathogens through nectaries, glandular hairs and other secretory structures is not uncommon. Bacteria have been shown to be able to enter undamaged nectar-secreting cells. Rain is usually associated with such an infection because the undiluted nectar frequently has an osmotic pressure too high to allow the survival of the pathogen. Dilution by rain enables the pathogen to survive and penetrate the surrounding tissue which has a lower osmotic pressure than the nectar. For example, a sugar concentration of over 10 per cent inhibits the growth of the fire blight bacterium *Erwinia amylovora* and this prevents infection through the nectary unless the nectar becomes diluted. Infection by fire blight bacteria has also been observed on undamaged leaves and is thought to take place through secretory hairs on the leaf.

Other suitable infection sites include all those parts of the plant with a thin cuticle and thin cell walls, such as the meristematic cells of a bud and the thin-walled motor cells of monocotyledon leaves, as in rice.

Despite the wide range of suitable infection sites on a plant it has often been observed that infections do not always start at such places. In fact there are many recorded instances of fungal spores germinating on the surface of a plant and the hyphae growing past apparently suitable infection sites, such as stomata, only to penetrate the host successfully at what is apparently a most unsuitable place.

4.7 Infection by animals

We have already considered transmission by animals (vectors) and this is sometimes followed by the animal actively infecting the new host. Not all vectors do this: bees, flies, etc, visiting fungal fruiting bodies to which they were attracted by a suitable secretion will transmit the infection but will do no more than deposit the spores on the surface of a new host. Similarly larger animals carrying spores or infectious particles on their coats will not be involved in the infection process. The sap-sucking and leaf-biting animals and the animals that burrow in plant tissues are likely to be involved in the infection process even if they are not involved in transmission. Frequently they are involved in both. The biting and burrowing animals simply introduce the inoculum on to a suitably damaged surface and the process of infection is just as if the spore had landed on a freshly wounded surface. Animals producing such effects include insects, nematodes, gastropods and mammals. It is not only herbivores that can cause infection in this way. Any animal which damages plants in its passing (such as crushing grass or pushing through under-growth) may introduce pathogens into newly-created wounds. Simply brushing against a leaf or stem can damage many thousands of hairs. Herbivores are, of course, by the nature of their feeding going to produce many more and far more serious wounds than non-herbivores, but the latter are still worth considering.

The sap sucking animals introduce the inoculum in a less obvious manner, there being little visible damage resulting from their feeding. They are however likely to have introduced fungal spores, bacteria or viruses deep into the tissues of the host, generally into the phloem. Sometimes the small amount of damage assists the host in limiting the pathogen as there is little wound trauma to assist the infection but in other instances the small amount of wounding means there is little host reaction until the pathogen is growing rapidly and has become well established in the host tissue.

4.8 Inoculum potential

Inoculum potential has been defined by one author as '. . . the energy of growth of a parasite available for infection of a host, at the surface of the organ to be infected . . .' It is implied that this is 'per unit area of the organ to be infected'. Inoculum potential is an important concept to consider in many situations, such as sanitation, field trials, plant breeding and fungicide testing. This concept raises a lot of questions. How much inoculum is required to infect a host plant? Is there an absolute minimum below which infection never occurs? Does more inoculum always increase the likelihood of infection? How is inoculum potential related to host resistance?

It has been shown many times that a single fungal spore or a single bacterium can cause infection. Inoculation of a large number of plants each with a single spore would, however, usually result in only a small percentage of the plants becoming infected. Even if only one plant had become infected out of a large number inoculated with single spores this would have demonstrated that a single spore is capable of causing infection under suitable conditions. If the number of spores or bacteria applied to each plant as inoculum is increased the percentage of plants to become infected also increases. Why should this be? There seem at first sight to be two possibilities. Firstly not all the spores or bacteria will have the same vigour, some will have larger food reserves than others and there will be considerable variation between the vigour of different types of propagules. Even if the spores or bacteria originate from a culture grown from a single spore or bacterium there is still the likelihood that there will be vegetative genetic variation between them. Secondly a large number of spores cannot be put on exactly the same place, and they will cover a considerable area of the surface of the host tissue. Some may be near epidermal hairs or near a stoma, whilst others may be in a region with a thick cuticle overlaying lignified or thickened cells. In other words the suitability of their position as an infection site will vary.

An illustration of the importance of nutrient reserves is the finding that although the percentage germination of *Botrytis fabae* spores in distilled water remained the same during a period of aging the infectivity fell to one tenth the original when the spores were twenty-five days old and to one hundredth the original when the spores were thirty-five days old. Infectivity of the aging spores could be restored by suspending them in 4·5 per cent sucrose. This is presumably because it restores the carbohydrate reserves that had become so depleted during aging that although the spores could germinate they were unable to overcome the host resistance and establish an infection. Both the age of the spore and the availability of carbohydrate in the environment of the spore are of great importance in the potential infectivity of an inoculum. Similarly the nutritional status of the fungus producing the spores is important. When *Fusarium roseum* was grown on culture media of high and low carbohydrate content (Czapek-Dox medium full strength and diluted tenfold) it was found that the conidia produced by the fungus grown on high nutrient levels were far more infectious than those from the fungus grown on low nutrient levels but their germination on artificial medium was found to be the same. So the nutritional status of the parent fungus also causes variation in the infectivity of fungal spores.

In some instance it has not been found possible to infect a host with a single spore or bacterium, more than one being required. This may be due to low infectivity of the spores and a paucity of suitable infection sites making the likelihood of a single spore achieving infection very low. However this requirement for more than one spore to achieve infection has led to the idea that under certain conditions there is a synergistic interaction between infecting propagules. Let us consider how two separate spores could interact during germination to cause an infection that neither alone could produce. Let us suppose that one spore germinates and penetrates the epidermis but is then killed by host reaction. A second spore germinating in the same region will have its growth and vigour (infectivity) increased by absorption of the nutrients that have leaked out of the damaged epidermal cell. It is then able to overcome successfully the host reaction and establish an infection in the surrounding cells. It is possible to visualize a situation where more than two spores would be required to overcome the host's resistance and for a generalized invasion of the tissue to occur. It may be that even before penetration there is a synergistic action, in that a large number of spores germinating together may alter the micro-environment, eg by altering the pH.

In most naturally occurring infections it is unlikely that two wind-borne spores would land nearly simultaneously sufficiently close together to have the sort of effect discussed. It is even more unlikely that there would be large enough numbers to affect the micro-environment. However if the spores were transported by a vector or were liberated together in a sporangium, or stuck together by mucilage, then interaction between them becomes quite possible.

Although there is no conclusive proof there does seem to be considerable evidence that on certain occasions there are synergistic interactions between spores. It has been found that to cause infection of cereal grains with bunt (*Tilletia caries*) a minimum of 500 spores per grain were required, and to achieve infection of 50 per cent of the grain 90 000 spores per grain were required. This could well be an example of synergism as the *Tilletia* spores are shed in sticky masses.

So far we have only considered spores and bacteria as units of infection because these can be readily counted and quantified. Mycelium of a fungus is of course also infectious but it is more difficult to measure the amount of mycelium present and so estimate its infectivity. Should one measure just the mycelium that is in contact with the host or does the mycelium away from the host but continuous with the hyphae at the surface of the host also have an effect? As nutrients can be transported through the fungal hyphae for some distance to support the mycelium in contact with the host, that portion away from the potential host must be considered. The further away it is the further nutrients will have to be transported and therefore the less effect it will have. It has been found that the importance of the mycelium is inversely proportional to the distance it is away from the host being attacked. The constants for these proportionalities will vary with the pathogen, its vigour in the old host and the resistance of the new host. *Armillaria mellea* has specialized vegetative infecting structures called rhizomorphs, which are aggregates of a large number of hyphae. These can grow over considerable distances (several metres) from an infected tree stump to a healthy tree. Here is an example of the inoculum potential (i.e. its ability to infect another tree) being inversely proportional to the distance from the original infection. The infectivity probably depends on the rate of growth of the invading hyphae and an exogenous supply of nutrients increases its growth rate and its infectivity. This is another manifestation of the effect noted with spores that the higher the nutritional status of the infecting structure the greater its infectivity.

The characteristics we have considered for mycelia and rhizomorpha also apply to sclerotia and other vegetative resting structures. The larger their food reserves the greater their inoculum potential. This presents a problem from the point of view of the pathogen. The larger the sclerotium or vegetative structure the greater its food reserve and hence the greater its inoculum potential but the fewer the number of propagules that can be produced from a given amount of substrate. It is not quite disperal versus inoculum potential because a large sclerotium can produce and support a far greater length of mycelium and thereby have a greater circle of invasion than would a small sclerotium. One can generally say that in a situation where there is a high density of hosts with considerable resistance to invasion, then the pathogens colonizing this ecological niche are likely to produce larger sclerotia with higher inoculum potential than pathogens with scattered, easily-infected hosts. (Scattered resistant hosts are unlikely to become infected.) Some pathogens (eg *Claviceps purpurea*) have it both ways, producing large sclerotia which produce large numbers of tiny spores.

So far, whilst considering inoculum potential, we have dealt with inoculation with an organism from a pure culture in the absence of any competing organisms. This is rarely the case in nature where there will be strong competition between micro-organisms prior to infection. This will of course influence the inoculum potential of the organism being investigated and might give quite different results to inoculation with a pure culture under sterile conditions. Several workers have found that they were quite unable artificially to infect wheat with 'take all' (*Ophiobolus graminis*) in unsterile soil whatever the dose of ascospores applied, whereas in sterile soil they were easily able to infect plants with low spore concentrations. The inability to infect in unsterile soil being almost certainly due to the inability of the pathogen to compete with the other rhizosphere flora.

In other instances the reverse can take place. As previously mentioned, organisms which are only weakly parasitic are unable to colonize healthy plant tissue but are able to colonize tissue already infected by a more vigorous parasite.

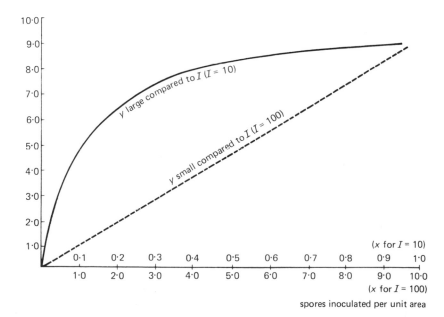

Figure 4.5 Showing the effect of increasing areas of diseased tissue on the number of spores required to cause a given increase in infection.

$$x = \frac{Ky}{I - y}$$

where; x = spores per unit area
y = percentage of host tissue infected
I = area of host tissue potentially infectable

4.9   The mathematics of synergism and independent action of propagules

If there is independent action of the propagules, i.e. no synergism, then $kx = y$, where x = the number of propagules, y = the number of infections and k = the fraction of propagules viable under the particular experimental conditions. This assumes an unlimited number of infection sites, but the number of infection sites will in fact be limited even if they are each one epidermal cell (in this instance there would be so many potential infection sites that under most conditions they could be considered unlimited). If however the infection sites were stomata, or some other region of limited extent, then the number of potential sites could well be sufficiently small to require consideration.

Then $y = (kx) \, k' \, (I - y)$

where; I = the number of potential infection sites and k' = the fraction of plant area that is infectable.

The two constants k and k' can be combined into one constant K;

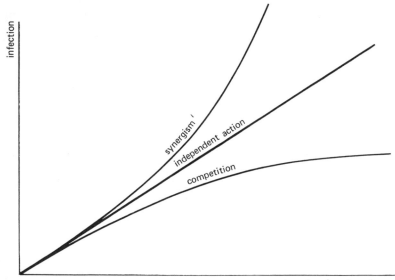

Figure 4.6 Where the spores act entirely independently there is a linear relationship between the number of spores in the inoculum and the amount of infection resulting. Where there is synergism an increase in the number of spores causes a more than proportional increase in the amount of infection, and where there is competition increasing the inoculum causes a less than proportional increase in infection. Interference may be caused in a number of ways, such as the inhibition of germinating spores by the metabolic by-products of those already germinated, or by a limited number of infectable sites on the host. It can be seen that synergism and interference are at a minimum when the number of spores is small and if the number is small enough the results approximate to the direct proportionality that occur with independent action.

then $y = Kx(I - y)$

so $Kx = \dfrac{y}{I - y}$

When y is small compared with I, $(I - y)$ approximates to I so that as we have seen $y = Kx$. As y gets larger $(I - y)$ can no longer be approximated to I which is why it has to be considered and the full formula used.

If y were small compared to I and there were synergistic interaction between the propagules there would, at least over some of the range of x, be a positive increase in the slope (i.e. the graph would curve up). If there were no interaction the graph would have a straight line, and if there were competition it would curve over (i.e. a negative increase in slope), giving a graph similar to that with y being large compared to I. If y were large compared to I, and there were synergism at the same time, it would be very difficult to separate these two effects because synergism would merely lift the curve and give the impression that there were a larger number of possible infection sites than were in fact present. It is possible that there might be a slight indication of this effect by the curve having a positive increase in slope at the foot and a negative increase at the head, i.e. having a somewhat sigmoid form.

From this it can be seen that if the interaction of spores or propagules is to be investigated the number of infections produced should be small in comparison with the number of infection sites. As it is rare that the number of infection sites is known it is wise to use the lowest spore numbers practicable in the particular experimental conditions used.

Chapter 5
Defence mechanisms

Despite the production of large amounts of inoculum by plant pathogens the majority of plants remain healthy. Infection is the exception rather than the rule. This clearly shows that most plants are resistant to practically all the pathogens with which they come into contact. There are however many examples where one member of a pair of closely related plants is resistant whilst the other is susceptible to a pathogen. This is of enormous interest to plant pathologists in that it enables studies to be made of the resistance mechanism, and is also of vital importance in the breeding of disease-resistant crop plants. The mechanisms of disease resistance are many and varied but they can be considered under a number of headings.

5.1 Preformed histological barriers

Penetration of a pathogen is frequently through the cuticle and epidermis, so that the structure of these tissues is important in the ease with which the initial infection can take place.

Table 5.1 *The relationship of susceptibility (to black stem rust) to the thickness of the outer epidermal wall and cuticle, and the age of the plants.* (after Melander and Craigie, 1927)

Susceptibility to black stem rust		Thickness of outer epidermal wall and cuticle	
(P. gramminis)	Species	Mature leaves (in μ)	2·3 day old leaves (in μ)
Highly susceptible	Berberis canadensis	1·29	0·88
	Berberis dictoyophylla	1·80	0·82
	Berberis vulgaris	1·87	1·10
Slightly susceptible	Berberis brachypodia	2·56	1·43
	Berberis lycium	3·41	1·23
	Berberis pruinosa	2·20	1·16
Resistant	Berberis thunbergii	2·44	1·57
	Oclostemon repens	3·01	1·75

It can be seen in the case of the infection of barberry by black stem rust that there is a trend in which the thicker the outer epidermal wall the more resistant the plant is to the rust. There is not however a perfect correlation between the wall thickness in a mature leaf and its resistance to infection (eg *Berberis lycium* would be expected to be resistant), but there is a better correlation between resistance and the epidermal wall thickness in young leaves. The increase in thickness of the outer epidermal wall as the leaves mature makes it unlikely that infection would occur through the mature leaves when young leaves are available with much thinner cell walls. Here is an example where one reason for a change in susceptibility with age is obvious. Clearly, when considering resistance, care must be taken to observe the correct developmental stage of the host. Thickness does not necessarily reflect the toughness of this outer layer and some measure of resistance to mechanical penetration could well give a truer indication of the part played by the outer epidermal wall and cuticle in resisting infection.

If this outer layer is the major difference between resistant and susceptible species it would be expected that wounding of resistant species and the inoculation of the underlying tissues would result in a susceptible reaction. This means the plant would show little resistance to infection introduced by biting and sucking insects, or by grazing and pruning wounds. In a number of instances puncture of the outer epidermal wall greatly reduces the resistance of a plant to infection, especially by unspecialized parasites, but generally other resistance mechanisms halt the growth of a would-be pathogen after penetration has occurred.

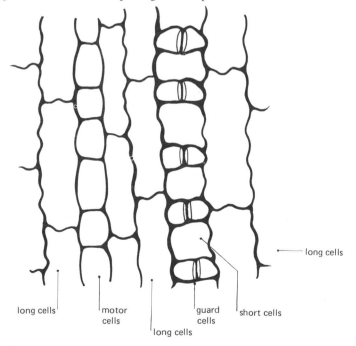

long cells

long cells   motor
            cells
                    long cells   guard   short cells
                                  cells

Figure 5.1. Surface view of rice epidermis showing the different cell types.
(based on drawings by Yoshi, 1936)

Calculations on the rôle of the outer cell wall and cuticle in resisting infection assume that other factors affecting the resistance of the test plants are the same. Certainly this is not always the case when one is comparing different species, so it is perhaps more relevant to compare different types of cell on the same plant where at least one is comparing genetically similar entities. This is not to say their *physiology* is identical, since they demonstrate different phenotypes, but this is probably the best comparison that one can make.

The blast disease of rice caused by the fungus *Piricularia oryzae* has been studied in connection with penetration and cell wall thickness. It was found that the pathogen usually entered through the motor cells, these being the hinge cells which cause the rice leaf to fold up under conditions of high transpiration so that the stomata are not exposed.

Table 5.2 *Site of penetration hyphae and appressoria of Piricularia on Komeji Rice.* (after Ito and Shimida, 1937)

Epidermal cell type	Percentage appressoria	Percentage penetration
Motor cells	53·0	63·7
Long cells	16·2	7·9
Short cells	12·8	6·1
Hairs	1·7	—
Stomata	6·0	—
All others, including middle lamella between cells	10·3	22·3

It was also noted that the outer walls of the motor cells remained in a pectino-cellulosic state long after the other cells had become lignified and it was considered

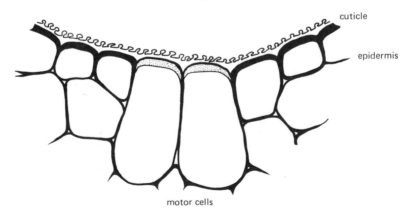

Figure 5.2 Transverse section of rice epidermis showing motor cells with their outer walls composed of cellulose and pectin and not liquified as are the outer walls of the other epidermal cells.
(based on drawings by Yoshii, 1936)

that the difference in penetration was due to the difference in mechanical toughness of this outer wall. As so often happens these results are not clear cut. Later workers have found lower concentrations of the fungitoxic compound chlorogenic acid in the motor cells than in the other epidermal cells and this could be the cause of the motor cells' lower resistance to penetration. It is not altogether clear that the chlorogenic acid content of a cell protoplast would have any effect on penetration of the cell wall, but it certainly would alter the survival of the fungus after infection as well as its ability to colonize other cells.

It is well established that the silicic acid content of rice leaves is inversely proportional to the incidence of disease. Silicic acid is usually deposited in large amounts in the epidermal cell walls and under favourable conditions it is also deposited in the walls of the motor cells. Planting young rice in flooded soil is found to increase the deposition of silicic acid and this is the cultural practice which has been found empirically over hundreds of years to reduce the incidence of disease. Low temperatures reduce the thickness and silicic acid content of epidermal cell walls and it is perhaps this that accounts for the sudden outbreaks of rice blast and *Helminthosporium* blight after a period of cold weather.

Berberis and rice are not the only plants to show the effect of epidermal toughness on disease resistance. Rust-susceptible flax has in general a poorly-defined cuticle and frequently has no hypodermis. Resistant strains on the other hand have a well-developed cuticle and hypodermis. In some strains the epidermal layer is so tough that the fungus is unable to rupture it and shed its uredospores. Although this may not alter the resistance of an individual plant it would certainly make the population more resistant to rust infection by reducing the amount of inoculum being produced.

Considering the number of pathogens that gain access to a plant through the stomata it is surprising how little the structure of the stomata affects a plant's susceptibility to all but bacterial infections. The susceptibility to bacterial infection is altered by the shape of the stomata, structures preventing the entrance of water also stop the ingress of bacteria. In the grapefruit *Citrus grandis* the stomata have broad oval lips allowing the easy penetration of water, whereas the mandarin organge (*Citrus nobalis* var. *Szinkum*) has an extremely narrow entrance to the stomata with high lips to the guard cells excluding practically all water. As expected, grapefruit is highly susceptible to the canker caused by *Pseudomonas citri* and the mandarin orange is highly resistant.

Stomatal behaviour is probably more important than structure in disease resistance. The resistance of some wheat varieties to black stem rust (*Puccinia graminis*) seems to be due to the fact that the stomata open very late in the morning. The relationship of this to disease resistance may not be clear until one considers the germinating uredospore growing through the dew on the leaf surface unable to penetrate the closed stomata. By the time the stomata open the dew will have evaporated and the delicate germ tube is likely to have been killed by desiccation. Closed stomata, however, offer no barrier to *Puccinia triticina*, the cause of brown leaf rust of wheat, and therefore varieties resistant to black stem rust are just as susceptible as any to brown leaf rust infections.

Not only does the time of day and water availability affect stomatal opening, the age of the tissue is also frequently important. *Cercospora beticola* is another pathogen that can only enter through open stomata. Young sugar beet leaves are resistant because stomatal movement has not started, mature leaves are severely attacked, especially when there is an abundant water supply (wilted leaves not being infected), and old leaves are resistant because stomatal opening has become very feeble.

Lenticels can also provide an entry for pathogenic organisms but this does not often happen in mature, heavily suberized lenticels. Immature lenticels which are only lightly suberized can and do provide easy access for infection. The cells of young lenticels are generally round and loosely arranged with fairly large intracellular air spaces. When the filamentous bacterium *Actinomyces scabies* attacks the young lenticels of a potato and causes potato scab it does so by growing down through these intracellular air spaces. The meristomatic cells of the lenticel are stimulated to rapid division and there is some attempt at cork formation to isolate the infection. Isolation is rarely successful as the fungus inhibits the suberization of the cork and the resulting mass of unsuberized cells produced in response to the infection gives the characteristic scabby lesion on the potato.

Normal mature lenticels have a cork cambium (phellogen) which is connected to the periderm of the stem. The cells produced by the cork cambium are closely packed, cuboid and heavily suberized forming a barrier to infection. Nearer the surface these cells become jumbled by the continuous growth of the cork cambium beneath them leaving many air spaces which harbour a wide variety of micro-organisms, many of which are potential pathogens. The lenticels of healthy mulberry trees contain the mycelia of a wide range of organisms, 30-60 per cent of which are pathogenic. Among these are *Diaporthe nomurai*, the cause of mulberry blight, and *Gibberella lateritium*, the cause of bud blight. Under normal conditions the cork prevents entry of the pathogens but under unfavourable conditions the pathogens can penetrate this layer. For example, *Diaporthe* causes blight when the trees are buried under heavy snow for a long period causing them to lose vigour, possibly due to anaerobic respiration. It is interesting to note that resistance to mulberry blight is inversely proportional to the fungal population of the lenticels. The rough-structured lenticels have more fungi than those with a close structure but the fungi in the rough-structured lenticels are unable to penetrate the cork whereas the fungi in the close-structured lenticels frequently do so. The mulberry trees with rough lenticels are therefore more resistant to blight than those with close textured lenticels. This may be due to the lower inoculum potential of fungi in the highly competitive environment of the rough lenticel and the increase in susceptibility under unfavourable conditions may in fact be due to a decline in the lenticel population, reducing competition for the pathogenic species.

Tissue within the body of plant which mechanically limits the spread of a pathogen is frequently called barricade tissue. This is often composed of lignified sclerenchyma cells and sometimes of suberized cork cells. The effect of such barriers is sometimes seen in leaf infections where the spread of the pathogen is limited by the lignified xylem elements in the leaf veins. In net-veined leaves this results in

localized patches of infection, and in parallel-veined leaves in long stripes of
infected tissue.

Just beneath the epidermis of wheat stems is a band of photosynthetic
collenchyma (chlorenchyma) which is interrupted by bands of lignified
sclerenchyma. In some wheat varieties, such as 'Little Club', there is comparatively
little sclerenchyma and there is not much more in the varieties 'Marquis' and 'Jota'.
Sonem emmer, on the other hand, has large amounts of sclerenchyma in this region
and little photosynthetic collenchyma. As a result, *S. emmer* is far more resistant to
both *Puccinia triticina* and *P. graminis* than the other wheats and even when it does
become infected under epidemic conditions there is far less damage done to it than
to the other varieties. It is probable, however, that the resistance is not due solely
to these mechanical barriers and other mechanisms also contribute to its resistance.

5.2   Infection-induced barriers

The very process of infection stimulates the metabolism of most plant tissues and
much of this metabolic activity is directed towards limiting the extent of infection.
Many of these defence mechanisms involve the laying down of some sort of barrier.

One of the most common reactions of plant tissue to limit the growth of a
pathogen is the rejuvenation of a band of cells ahead of the infected zone and these
cells start to divide and produce a layer of suberized corky cells. The success of this
defence reaction depends on its speed and vigour. The vigour of the pathogen is
important in turn in determining the amount of time available to the plant to form
a suitable barrier. If there is insufficient time, or the barrier produced is inadequate,
the pathogen will breach it and invade the underlying tissues. Stimulation of the
host plant to produce the cork cambium might be due to metabolic by-products of
the pathogen or it could be due to materials liberated from mechanically wounded
cells diffusing ahead of the invading pathogen.

Cambial activity results in some instances in the sloughing off of necrotic and
infected cells thus reducing the inoculum potential for further penetration. Even if
this does not happen a mass of necrotic infected and corky cells is often pushed up
away from the surface and away from the region of active invasion. In some
instances the laying down of a corky barrier cuts off the supply of water to the
outer layers so that they dry out and collapse to form a dry necrotic depression.
These different reactions result in characteristic types of scab lesion. Scab diseases
are caused by a number of different fungi and affect a large number of plants,
particularly storage tissues in the stem, root or fruit.

Suberization does not always result in typical scab lesions. When *Prunus
domesticata* is attacked by *Coccomyces prunophorae* the lesion on the leaf is
isolated by a ring of suberized cells running right through the leaf rather than in the
dish shape that is common in storage organs.

Sweet potato is a good example to study because different strains show different
reactions when attacked by *Helicobasidium mompa*. The fungus grows epiphytically
on the surface of the tuber for some time, producing a purplish felt-like mass of
rhizomorphs. During this time the hyphae penetrate the middle lamella of the

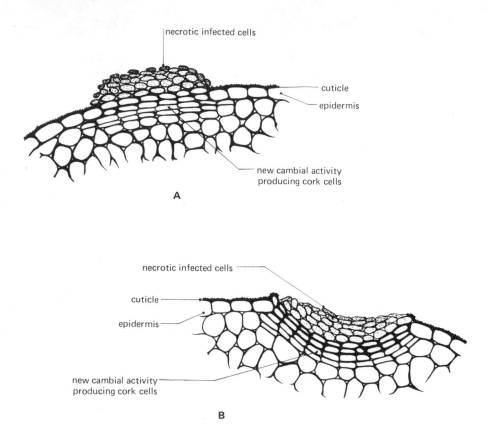

Figure 5.3 Two ways in which corky barricade tissue can be produced: **A** Near the surface pushing up the infected tissue which may become sloughed off, and **B** produced in a depression in which the infected tissue is isolated.

suberized cells that normally surround a sweet potato tuber. The penetration hyphae gradually produce an infection cushion which then grows out, penetrating the corky outer layers of the tuber. The tubers react in different ways depending on their variety and on the environmental conditions. The infection may be quickly halted by a layer of cork, or an appreciable amount of infection may have occurred before the fungus is stopped. Sometimes the fungus is not stopped and the brown rot symptoms spread throughout the tuber. On yet other occasions the production of large amounts of pectolytic enzymes by the pathogen causes the entire tissue of the tuber to collapse without browning.

The rate of cork formation is affected by temperature and humidity and this will, of course, alter the host's ability to resist attack. The cork layers are produced in sweet potato most rapidly at 30-35°C and at 90-95 per cent relative humidity. Humidity above this reduces the rate of cork formation and humidity below 80 per cent inhibits its formation completely. In the Irish potato the optimum temperature for cork formation is lower, lying between 21°C and 35°C, but the optimum relative humidity is similar. Soil moisture clearly influences the rate of cork forma-

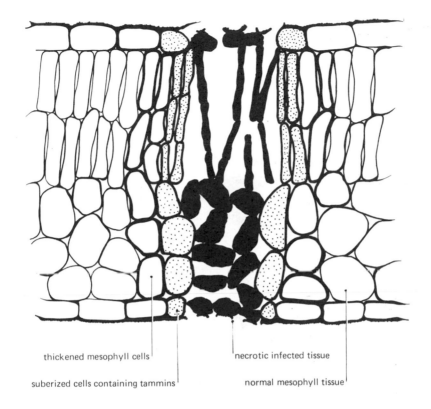

thickened mesophyll cells

suberized cells containing tammins

necrotic infected tissue

normal mesophyll tissue

Figure 5.4 Corky barricade tissue forming a cylindrical barrier right through a leaf thus isolating a core of infected tissue which may or may not be shed.

tion in underground tubers by altering the humidity but it can also alter the aeration. Waterlogged soils with anaerobic conditions inhibit the formation of cork as well as favouring the growth of pathogenic facultative anaerobes. This is why black leg of potato is so common in waterlogged soils.

Lignification can also occur as a means of limiting the growth of a pathogen but it is not nearly as effective as suberization. Lignified tissue is usually penetrated by the pathogen but it delays the progress of the disease and may give time for other defence mechanisms to be completed.

Another mechanism by which the invasion by a pathogen can be limited is the shedding of infected tissue, an effect particularly common in fruit and leaf infections. A band of cells, either directly round the infection or at the point of attachment of the organ, start to divide and produce thin-walled cells with large amounts of pectin in the walls and marked middle lamellae. The cell walls and middle lamellae break down so that there is no longer any connection between the infected and healthy tissues and the infected tissue is shed. The shape of this abscission layer varies; sometimes it is cup-shaped and on other occasions, parti-

The invasion hyphae are prevented from reaching the starchy parenchyma by the formation of a corky barrier layer.

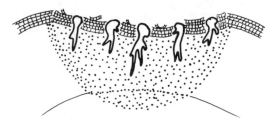

The infection reaches the starchy parenchyma but at about this time is halted by the formation of a cork. The infected tissue displays a brown rot appearance.

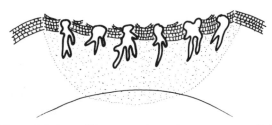

Rapid growth of the infection cushions causes the development of brown rot but no cork formation. The whole tuber therefore becomes infected.

Pectolytic activity of the fungus macerates the tissue without browning. No cork is formed so that the whole tuber is rapidly infected.

Figure 5.5 Four different reactions of sweet potato to *Helicobasidium mompa* depending on variety of potato and environmental conditions. (redrawn from Suzuki, 1957)

cularly in leaves, it is cylindrical, passing right through the leaf so that a disc of tissue is shed giving a shot-hole effect. With leaves and fruits which will be shed in autumn there is often already the beginnings of an abscission layer which is activated by infection so that the whole organ is shed. If the pathogen has not crossed the abscission region when the infected tissue is shed this is an extremely effective way for a plant to rid itself of a pathogen.

Chemicals can induce abscission in a similar way to bacterial and fungal infections. Dilute copper sulphate (0·01 Molar) can produce a response similar to infection. It has been noticed that peach trees planted under copper wires have a shot-hole effect produced as a result of the copper leached from the wire dripping on to the leaves.

It has been observed that leaves of almond (*Prunus amygdalinus*) attacked by *Cladosporium carpophilum* respond in different ways depending on the age of the tissue and the availability of water. Young leaves well supplied with water form an abscission layer round the infection producing perforations of the leaf but in older leaves, and in leaves not so well supplied with water, no abscission layer is formed and the cells around the infection become suberized instead.

Although shedding infected tissue is frequently successful in limiting the spread of the pathogen within the host it does not always control the life cycle of the pathogen. On twigs of cherry (*Prunus yedoensis*) attacked by witches broom caused by *Taphrina cerasi* small leaves are produced on the lower surface on which the fungus develops ascospores within a sorus. The sorus develops in a localized portion of the leaf, abscission cells are produced around it and finally the sorus is sloughed off but only after the ascospores have been shed and the life cycle of the pathogen perpetuated. The abscission layer seems to be produced more in response to the death of some of the leaf tissue than in response to the presence of the pathogen. This phenomenon could well be more general than we realize, the host responding not to the presence of the pathogen but to the damaged, dead and dying cells of its own tissue.

Infections of the vascular system can result in the production of tyloses. These are formed by vesicles extruded from neighbouring parenchyma cells through connecting pits into wood vessels or tracheids, thus blocking or partly blocking the vessels. There is no general agreement about the cause of tyloses and it seems likely that there are many stimuli that will cause a parenchyma cell to push out a vesicle in this way. It was initially suggested that tyloses are produced in response to the autolytic products of damaged cells, an idea based on the observation that tyloses are frequently found associated with amputation of tree branches and adjacent to wound areas. Other workers viewed these observations in a different light and suggested that it was the entry of air into xylem elements that triggered tylose production. As the result of further studies it was proposed that tyloses were produced in response to the toxic products of invaded cells and that these products were not translocated but affected a restricted region of the xylem. Fairly recent work has shown that tyloses in chestnut trees are produced in response to a toxin called diaporthin liberated by *Endothia parasitica*. Culture filtrates of a pathogen can induce wilting symptoms in cuttings of the host; for example, the culture filtrate of *Verticillium albo-atrum* will induce wilting of lucerne cuttings. These various hypotheses about the stimulation of tylose formation need not be in conflict. It is possible that different mechanisms operate in different host-pathogen combinations.

The function of tyloses is as unclear as their formation. They are very thin-walled so one would expect them to have little ability in isolating the fungal mycelium

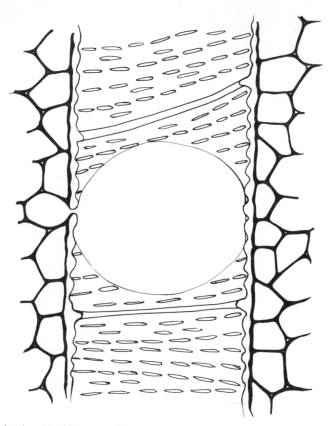

Figure 5.6 A tylose blocking a wood vessel. This is the result of one of the xylem parenchyma cells pushing out a thin-walled vesicle through a pit in the wall of the wood vessel.

which would grow straight through them, but surprisingly the growth of vascular wilt pathogens through a tylose has not been observed. It is possible that the sealing of xylem elements carrying metabolically toxic materials from whatever source, is advantageous to the host, since the blockage stops the materials reaching the leaves. Conidia have been observed moving in the transpiration stream of *Verticillium*-infected lucerne plants, and tyloses would limit this form of spread. In the xylem of sweet potato, tyloses are produced well ahead of the invading mycelium of *Fusarium oxysporum* so that such a sealing mechanism against both toxins and spores is possible.

The surprising infrequency with which tyloses seem to be penetrated could be due to the gums and resins which tend to be produced with them. The gums produced in response to wounding or infection have many of the characters of lignin and act as additional mechanical barriers to the pathogen. In the absence of tyloses being secreted into wood vessels, these gums and resins are also produced probably through half bordered pits in order to block the vessel in vascular infections in much the same way as tyloses. Gums and resins are found not only in the vascular system, but are also produced in response to localized infection and in some instances are important in limiting the spread of a lesion. These compounds resist penetration both by mechanical strength and by containing fungitoxic substances, largely polyphenol derivatives, that give them their characteristic brown colour.

The difference in the rate of development of a disease under different conditions is often the result of the suitability of these conditions for gum formation by the host. Conditions favouring gum formation restrict the spread of the pathogen. Similarly host varieties that rapidly produce large amounts of gums and resins in response to infection tend to have a fairly high resistance to the disease.

Table 5.3 *Reaction of cells in rice leaves related to susceptibility to blast caused by* Piricularia oryzae. *(after Kawamura and Ono, 1948)*

Variety	Speed of Reaction	Resin/Gum	Resistance/ Susceptibility
Kan Non Sen	rapid	abundant	highly resistant
Gin Nen	rapid	considerable	resistant
Ko Sen	slow	small	resistant
Kameji	slow	considerable	resistant
Kairyo Shinriki	slow	considerable	susceptible
Wase Asashi	slow	small	susceptible

It can be seen that resistance is usually related to the speed and abundance with which the gums and resins are produced. However it is obvious that other mechanisms are involved in resistance, particularly of species like 'Ko Sen' which reacts slowly to infection, produces only small amounts of gums and resins and yet is still fairly resistant to blast. It is also clear that other factors are affecting the resistance of 'Kameji' and 'Kairyo Shinriki': both react in a similar way with regard to gum and resin formation, but one is fairly resistant whilst the other is susceptible.

Isolation of the pathogen by gums and resins in parenchyma or mesophyll tissue is an effective defence mechanism with few side affects but occlusion of the vascular system will, of course, result in reduced translocation and in some cases wilting and premature leaf fall of the host. It may even result in death. The blockage of the vascular system usually occurs in perennials so that if it is successful in isolating the pathogen the new growth of the wood the following year will replace that lost by blockage and little permanent damage will have been done. In plants with perennial root stocks the damaged stems can be replaced by new adventitious shoots. However it is unusual for a pathogen to be killed or even totally contained in this way and the host reaction merely slows down the rate of spread, perhaps enough to allow the plant to set seed and complete its life cycle.

Attempts at cell wall penetration by some potential pathogens is resisted by the deposition of callose around the penetration hypha. The increase in thickness of the host cell wall has frequently been observed prior to penetration of the pathogen through the original cell wall, so that clearly a reaction is taking place before the pathogen reaches the protoplast of the cell. This reaction does not always deter the pathogen and penetration may still be successful. The walls of pea epidermis have

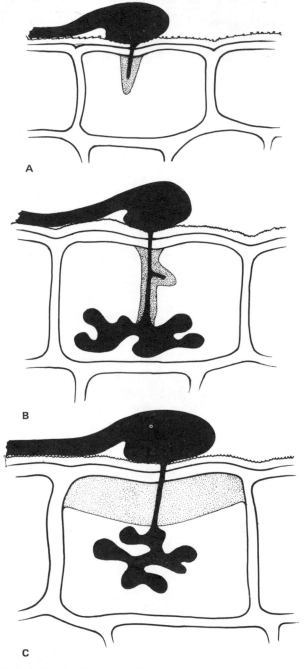

Figure 5.7 Three examples of callose deposition in response to penetration: **A** A callose deposit round the penetration peg which may or may not be successful in stopping infection; **B** where the callose deposit around the penetration peg has not stopped infection and the fungus has grown through it to produce a haustorium; **C** callose deposited across the entire outer wall of a cell in response to penetration which in this instance has not stopped the infection.

been seen to swell before the successful penetration by *Botrytis cinerea* and the pathogen has so little difficulty growing through this region it has been suggested that the swelling is a result of enzymic softening of the cell wall by the pathogen. In many cases however the production of callose does slow down the penetration of the pathogen and sometimes stops it altogether, particularly when the callosity becomes lignified (when it is called a lignituber). Tomatoes are totally resistant to the rice blast pathogen *Piricularia oryzae* at least partly because of the heavily lignified callose deposit around the penetration hyphae which completely stops the growth of the pathogen. A similar effect is seen in oats attacked by *Helminthosporium avenae* and in flax attacked by *Fusarium lini* but in these cases the lignituber is not always successful in halting the pathogen so that resistance is not complete. Despite these observations there does not seem to be a direct relationship between callose production and resistance, some of the most susceptible plants being the most prolific in their production of callose. Some workers think the callose production is one of a number of defence mechanisms operating at any one time and resistance or susceptibility depends on the sum of these reactions, while other workers consider the production of callose is part of a wound healing reaction and not directly related to resistance at all.

Callosities are induced not only by fungal pathogens since mechanical injury will often stimulate their production. Pricking an epidermal cell with a fine needle frequently results in callose formation almost indistinguishable from that produced in response to a pathogen. It would therefore seem quite likely that the stimulus for the production of callose is the mechanical penetration of a cell wall by any one of a variety of agents. In general callosities are formed most readily and most rapidly in young vigorous cells and more slowly in older cells. This probably reflects the rate of metabolic activity in these cells and the greater need for young, thin-walled cells to add an additional mechanical barrier to invasion.

Something which in the mature state looks very like a callosity is the deposition of cellulose over a fungal hypha that has already penetrated a cell. This may limit its growth and uptake of nutrients, thus reducing the seriousness of the infection. In the early stages it is easily distinguished from a callosity because the layer of cellulose is much thinner and looks like a thickened hyphal wall, but later it can build up to a considerable thickness so that the structure looks like a callose deposit. This process is perhaps best described as encapsulation and is also a defence reaction against a pathogen but one occurring at a later stage than callose production.

5.3   Pre-existing biochemical defences

The absence in a plant of essential compounds which the pathogen is unable to synthesize for itself cannot really be considered a defence mechanism but it is convenient to deal with it here because it does seem to be of importance in resistance and susceptibility of plants to some diseases. In virus infections, where the whole cellular metabolism is altered for the synthesis of new viral particles, a very special cellular make-up is required to enable the virus to be successful in this re-direction of the metabolism. It is not surprising, therefore, that viruses have a very limited host range, and most plants are not infected by most viruses. With bacterial and fungal pathogens the situation is not quite so obvious. Some totally obligate pathogens may have lost the genetic information for the production of a particular metabolic intermediary and this is provided by the limited number of plants which the pathogen is able to colonize. Clearly the requirement must be for something more sophisticated than that required by a simple nutritional mutant because most plants would provide such simple requirements.

That pathogens capable of growing on relatively simple media have a limited host range shows that even plants showing little in the way of histological barriers to infection have active defence mechanisms. These must be biochemical and physiological. Biochemical defences may be present all the time in a healthy plant or they may only be produced at a particular age or stage of development, thus accounting for variation in susceptibility with age. The production of compounds, inhibitory to the pathogen may also be enhanced or repressed by environmental conditions, accounting for environmental variation in susceptibility. These compounds act in a number of ways and at a number of sites. They may act on micro-organisms growing on or near the surface of the host, reducing their vigour and their ability to invade the host tissue. Alternatively the inhibitors may totally stop the growth of a particular organism on its surface either by killing the mycelium or more commonly by inhibiting spore germination. These anti-microbial compounds might act within the host to halt the growth of the pathogen following penetration and so prevent the establishment of an infection. On the other hand they may not be quite as effective as this, only slowing down the growth of the pathogen and allowing other defence mechanisms to be effective, or allowing the host to reach maturity and resistance.

The demonstration that a crude or purified extract of host tissue has anti-microbial properties does not prove that it is the mechanism of resistance. Before considering a substance as the cause of resistance it must satisfy the following criteria: (i) the substance must be present in those parts of the plant showing resistance. It is pointless to demonstrate the presence of a microbial inhibitor in the leaf when infection occurs through the root. A frequent and more subtle error on the same lines is the demonstration of an inhibitor in the organ under attack without showing it to be present in the tissue being invaded, and even the demonstration of an inhibitor within a tissue liable to microbial attack does not make it certain that it is involved in resistance. Resistance may be due to materials secreted on the surface of the organ where they inhibit the pathogen, while substances within the tissue may play no part at all. (ii) The substance must be in high enough concentration at the site of action to have the effect observed *in vitro*. The presence of high levels of nutrients found in plant tissues frequently reduces the sensitivity of a pathogen to an inhibitor and this too must be considered when investigating the importance of a possible anti-microbial compound. (iii) It must be shown that the inhibitor is available to the pathogen. The extraction procedures used may liberate substances from the vacuole and cell organelles which may in the normal course of events never come into contact with a pathogen growing between cells and pushing out haustoria into the cytoplasm. (iv) It should be demonstrated that the substance suspected of causing the resistance in resistant varieties is not present, or at least not available to the pathogen, in susceptible varieties. (v) It should be possible, at least in some diseases, to convert a susceptible specimen to a resistant one by treatment with the anti-microbial compound.

There is a wide range of substances acting on a pathogen even before penetration of the host is attempted. In 1957 it was found that the exudate from peas resistant to some of the races of *Fusarium oxysporum* f. *pisi* inhibited the germination of spores from the races to which the peas were resistant but had no effect on the races to which they were susceptible; i.e. the resistance appears to be due to inhibiting the germination of the pathogen spores. This correlation between resistance and inhibition of conidial germination is not conclusive, particularly as in this instance there is not total inhibition of germination, and those conidia that do germinate have perfectly normal germ tube and hyphal development. An effect such as this would, of course, reduce the effective concentration of spores, thus having the same initial effect as sanitation, as well as reducing the rate of increase of the pathogen. It has been suggested that it may be more relevant in this sort of experiment to use chlamydospores rather than conidia as the test spores because it has recently been shown that chlamydospores are frequently the more important form of inoculum in the soil.

There is evidence that amino acids and sugars leached out of seeds prior to germination affect their susceptibility to unspecialized pathogens such as *Pythium*. Those seeds with the larger amounts of exudate tend to be more susceptible than those with small amounts. This probably results from increased inoculum potential. As we have seen, such exudates can quite rapidly alter the microbial population round plant roots (the rhizosphere) so that it is less favourable to the pathogen; therefore stimulation of the pathogen by an exudate is only likely to occur early in the germination of a seed.

We have also commented on the stimulatory and inhibitory effects of root and leaf exudates but these have been fairly non-specific and do not really account for the susceptibility of a plant to some races of a pathogen and its resistance to others. However, a specific relationship has been shown in some instances. Washings from beet resistant to *Cercospora beticola* inhibit the germination of these spores more than washings from susceptible beet varieties. The best example of substances acting outside the host to resist a pathogen is the case of onions attacked by the smudge disease caused by *Colletotrichum circinans*. Following growth as a saprophyte on the dead outer scales the fungus attacks the fleshy living scales within. Varieties of onions in which the outer scales are coloured yellow or red are resistant whilst those with white scales are susceptible, and in those varieties of onions which are less coloured at the neck than at the base it is at the neck that penetration occurs. The pigments include flavones, probably as the glycosides, and were thought at one time to be the cause of resistance as similar compounds have been shown to be inhibitory to some fungi. If the outer scales are removed, however, infection of the fleshy inner scales occurs despite the fact that they contain the pigments. It has now been found that germination of *C. circinans* spores is inhibited by substances diffusing out of the dead coloured scales which do not diffuse out of the live scales. The compounds have been identified as catechol and protocatechuic acid, and the pigmentation is purely coincidental.

Perhaps more important than the substances diffusing out of plants and inhibiting attack by pathogens are substances remaining within the host and stopping the growth of pathogens after penetration. Often the anti-microbial substances found to have diffused out of plants are also active in stopping the growth of pathogens within the host but there are also many compounds within a plant which are important in resisting infection and which are never found liberated at the surface. The very wide range of chemicals involved in this way includes phenols, alkaloids, glycosides, amino acids, proteins, carbohydrates and members of many other groups of compounds.

Phenols have for many years been investigated as the source of disease resistance in plants. It has been claimed that the phenols, the quinoles formed from them by oxidation, and the products of further oxidation, are largely responsible for the resistance of higher plants to infection. Many other pathways have now been shown to be involved in disease resistance but the phenols are undoubtedly of enormous importance. A considerable time ago research workers showed that black stem rust (*Puccinia graminis f. tritici*) attacks wheat varieties with low phenol content more frequently and causes more damage than to varieties with high phenol content. Injecting susceptible wheat varieties with a dilute solution of vanillin or catechol has been found to increase their resistance to the disease. Despite these findings it is unfortunately not possible on an agricultural scale to increase the resistance of wheat to black stem rust by application of polyphenols because the situation is

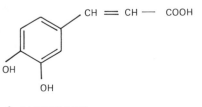

A CAFFEIC ACID

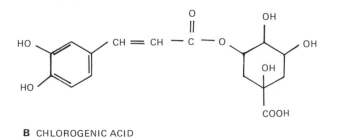

B CHLOROGENIC ACID

Figure 5.8 Two phenolic fungitoxic compounds present in many plant. **A** Caffeic acid **B** Chlorogenic acid.

modified by a whole host of other resistance and susceptibility mechanisms making the relationship of polyphenol to resistance most complex.

The relationship between phenol content of the tissue and resistance to infection in the common scab of potatoes caused by *Streptomyces scabies* appears to be simpler. Tubers of resistant varieties have a consistently higher chlorogenic acid content than tubers susceptible to the disease. In resistant varieties histochemical staining techniques show the highest concentration of chlorogenic acid around lenticels and wounds through which the pathogen might enter. In tests *in vitro* caffeic acid and catechol as well as chlorogenic acid were found to inhibit the growth of the pathogen at low concentrations. Their toxicity increased as the pH of the medium increased from 6·0 to 8·8. This is probably related to the high rate of oxidation of phenols under alkaline conditions as there is evidence that the oxidation products of phenols and polyphenols are more inhibitory to micro-organisms than is phenol itself.

A considerable amount of work has been done on the polyphenol metabolism of potatoes infected with *Verticillium albo-atrum,* a vascular wilt disease. The concentration of *o*-dihydroxyphenols tends to be higher in the underground parts of resistant varieties than in susceptible varieties and more significantly the concentration falls more rapidly after planting in the susceptible varieties than in the resistant ones. In the field the rapid development of the disease in the susceptible varieties coincides with the fall in concentration of *o*-dihydroxyphenols. Development of the disease can be delayed by defoliation and by the application of growth retardants, both treatments causing an increase in phenol concentration or at least stopping the decline in concentration that would normally occur.

Table 5.4 *Chlorogenic acid content of potato leaves and roots.* (Lee and Le Tourneau, 1958)

Variety	Resistance to Verticillium	Chlorogenic acid Percentage of fresh wt	
		leaves	roots
Popular	extremely resistant	0·17	0·08
41956	very resistant	0·25	0·07
Great Scott	resistant	0·12	0·11
Early Gem	susceptible	0·17	0·01
Kennebec	susceptible	0·14	0·05
Russet Burbank	susceptible	0·18	0·01
Bliss Triumph	very susceptible	0·08	0·03

Table 5.5 *Chlorogenic acid content of tubers following planting.* (McLean et al, 1961)

	Chlorogenic acid content (relative units) at the number of days indicated after planting				
Susceptible varieties	108	122	155	180	200
Early Gem	8	8	5	4	3
Russett Burbank	8	4	5	4	3
Kennebec	7	7	4	2	—
AVERAGE	7·7	6·3	4·7	3·3	2·0
Resistant varieties					
Populair	11	12	13	11	10
41956	17	15	16	12	9
Great Scott	23	23	18	11	6
AVERAGE	17·0	16·7	15·7	11·3	8·3

There are similar correlations between polyphenol content of host tissues and its resistance to infection by a number of diseases. These are found in resistant and susceptible varieties of apple to apple scab (*Venturia inaequalis*), and pear to pear scab (*V. pirina*), in potatoes to *Phytophthora infestans*, lima beans to *P. phaseoli*, sugar beet to *Cercospora beticola,* rice to *Piricularia oryzae,* and many more. In none of these cases, however, have all the criteria previously listed been satisfied so that although the implication of polyphenol involvement in resistance is exceedingly strong it is not absolutely certain.

Although polyphenols appear to be important in disease resistance of a wide range of plants there is no indication that it is universal as a resistance mechanism, and in fact there are a few isolated instances where polyphenols stimulate the growth of a pathogen. The polyphenol gallic acid, which is present in the capsule of *Riccinus communis* (a bryophyte), stimulates the growth of the pathogen *Botryotinia ricini* which attacks the capsule. Gallic acid stimulates both the germination of conidia and the growth and sporulation of the mycelium at concentrations less than 0·1 per cent.

Specific alkaloids are often found in particular species, or groups of species and because of their marked biological activity are certainly candidates for a rôle in the resistance mechanism of the organisms in which they are found. In some cases there are good grounds for such a suggestion. For instance, *Phymatotrichum omnivorum* which, as its name implies, attacks a wide range of plants by destroying their roots, is unable to attack *Mahonia trifoliata* and *M. surasii* both of which contain the alkaloid berberine in concentrations far higher than that needed to inhibit totally the growth of *P. omnivorum* in artificial culture. There are indications that alkaloids are also involved in the resistance of other species to this pathogen; *Sanguinaria*

Plate 1   Symptoms

A   *Taphrina deformans* causing distortion, loss of chlorophyll and formation of red pigments in the infected areas of almond leaves. Heavy infections result in leaf fall and regrowth.

B   *Microsphaera berberidis* forming an extensive growth on the surface of mahonia leaves.

C   Black spot on roses caused by *Diplocarpon rosae* in which chlorosis of large areas of the leaves occur as well as the black necrotic regions. Heavy infections cause leaf fall and regrowth.

D   Local lesions produced by inoculating french bean leaves with tobacco necrosis virus. The infection does not become systemic in this host.

A

C

D

B

Plate 2   Rusts

A   A photomicrograph of an aecium of *Uromyces dactylidis* on *Ranunculus*.
B   Aecia of *Puccinia lagenophora* on the stem of groundsel. Note the enlargement and distortion of the host tissue.
C   Teliospores of *Puccinia sauveolens* on thistle.
D   Spermagonium of *Puccinia sauveolens* on thistle.
E   *Puccinia sauveolens* infecting a thistle plant. Note distortion of stem.

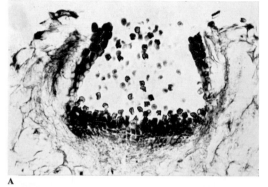

A

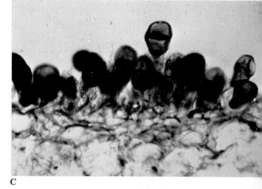

C

B

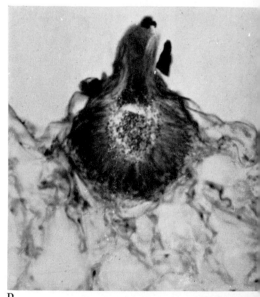

D

E

Plate 3   Dutch Elm disease

A   The infection spreading back from the tips of branches where the initial inoculation occurred during the feeding of the adult *Scolytus* beetle.
B   The mycelium of the pathogen (*Ceratocystus ulmi*) growing saprophytically from the cut end of an infected elm log.
C   The burrows made by the larvae of the beetle *Scolytus* whilst feeding beneath the bark of an infected branch. From here the adults will carry the fungus to the tips of healthy trees where they go to feed, thus completing the life cycle of the fungus.
D   Dead elm trees (centre right) from which the infection has spread to the neighbouring trees (centre left) which are showing the first symptoms of dutch elm disease, yellowing of the upper and outermost leaves.

C

D

Plate 4

A Haustoria of *Peronospora parasitica* in cells of wallflower.
B An ergot formed by *Claviceps purpurea* in place of a seed on an infected spike of perennial rye grass.
C Smut spores produced by *Ustilago nuda* in place of seeds in an infected barley plant.
D Showing that infection with *Ustilago nuda* causes the barley to grow taller and to produce its flower spike earlier so that the spores of the pathogen are carried up higher than the surrounding plants from where the spores are dispersed by the wind to the other plants infecting their flowers.
E The systemic nature of the smut infection is clearly demonstrated by the occasional production of spores in the vegetative tissue of the host as well as in the seed.

A

B

C

D

E

Plate 5   Wounding

A   Gums and resins deposited in the vessels of an oak tree following pruning wounds as a defence against infection. (Scanning electron micrograph, V. Monk, University of Surrey.)
B   A section through a large tylose produced in an oak wood vessel as the result of a pruning wound. (Scanning electron micrograph, V. Monk, University of Surrey.)
C   A number of small tyloses produced in an oak wood vessel as a result of a pruning wound. (Scanning electron micrograph, J. Sargent, University of Surrey.)
D   Regrowth around a major wound which has not been successful in stopping infection by a rot pathogen.
E   The effect of micro-wounds produced by adding a fine abrasive (kieselguhr) to the inoculum of tobacco necrosis virus applied to french bean leaves. Plants on the left have fewer lesions than those on the right on which the remains of the kieselguhr are still visible as a white powder.

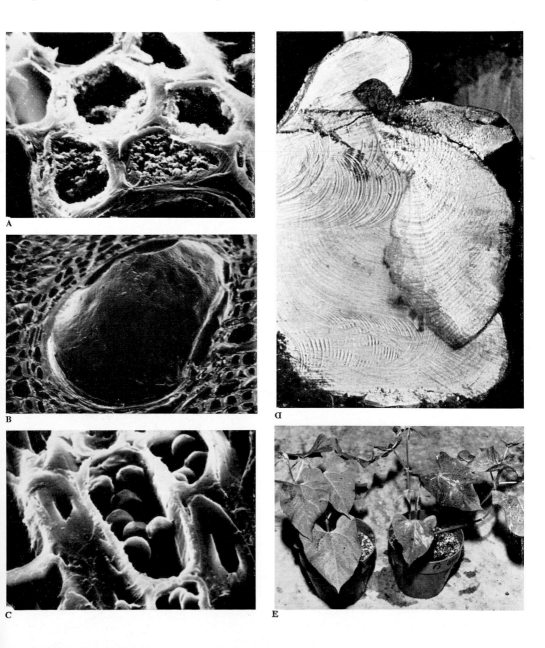

Plate 6   Penetration

Transmission electron micrographs of *Colletotrichum graminicola* penetrating maize cell walls.

A   A section through a hypha with the penetration just starting to be produced. Even at this early stage a papilla of host cell reaction products is visible on the opposite side of the cell wall to the penetration peg.

B   A section through a hypha with the penetration peg just breaching the host cell wall. Note increased papilla of host reaction products.

C   A section through a hypha successfully penetrating a host cell wall and starting to form a haustorium. The plasma lemma (PL) becomes invaginated but is not ruptured and there is still a large papilla (P) of host reaction products.

D   A section through a penetration peg that has not yet successfully breached the host cell wall due to attempting to penetrate at the junction of cells and partly due to the large amount of host reaction products. (Electron micrographs, D. J. Politis & H. Wheeler, University of Kentucky.)

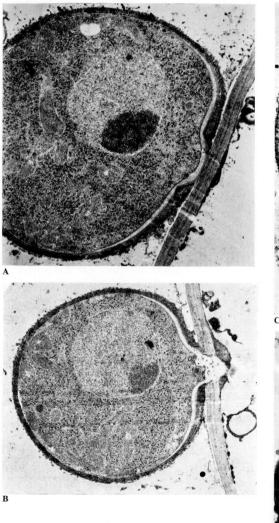

A

B

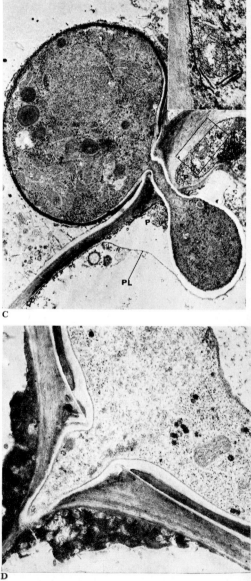

C

D

Plate 7 Powdery Mildew

Scanning electron micrographs of the surface of hop leaves infected with *Sphaerotheca humuli*.

A Spores germinating.
B Extension of mycelial growth.
C Fruiting bodies starting to be produced.
D Large numbers of chains of conidia growing up from the surface of the leaf. (The large structures on the surface of the leaves are epidermal hairs.) (Scanning electron micrographs, D. Royle & A. de S. Liyanage, Wye College, University of London.)

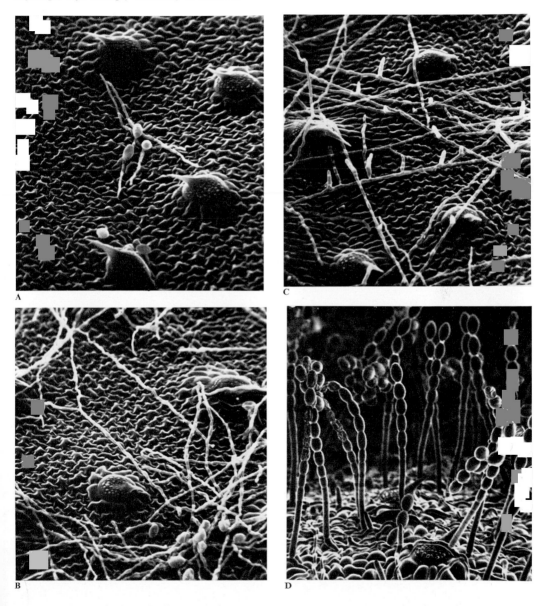

A

C

B

D

Plate 8 Penetration stimulus

A Scanning electron micrograph of several zoospores of *Pseudoperonospora viticola* achieving penetration through an open stomata on a hop leaf.

B Groups of zoospores of *Pseudoperonospora viticola* aggregated around open stomata. Aggregation such as seen in (a) and (b) does not occur around closed stomata. To investigate this phenomenon more closely moulds of leaf surfaces with closed and open stomata were made from which plastic replicas were cast.

C Plastic model of leaf surface with stomata open with which zoospores of *Pseudoperonospora humuli* are associated.

D Plastic model of leaf surface with stomata closed. Little association of germinating zoospores with stomata.

These results indicate a tactile response on the part of the zoospores to the open stomata. (Micrographs D. J. Royle & G. G. Thomas, Wye College, University of London.)

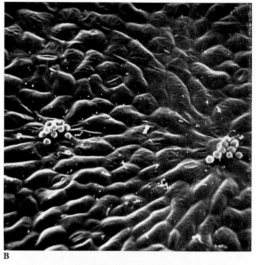

B

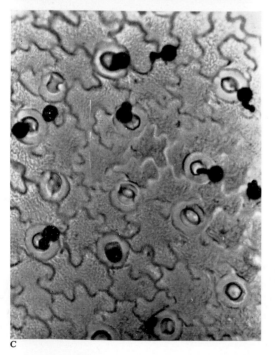

C

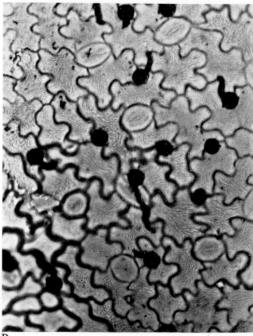

D

Plate 9   Defence mechanisms

A   The skin of the lemon forms a preformed barrier to infection by soft rot fungi. Once breached the pathogen meets relatively little resistance to further growth. The host tissue is macerated in advance of the growing mycellium by pectolytic enzymes liberated by the fungus.

B   Some plant tissues contain preformed antifungal compounds. In this case the diffusate from the homogenized garlic (right) has halted the growth of *Pythium* from the inoculum block (left) and has caused morphological changes to the mycellium (centre).

C   Young stomata before they open for the first time are covered with a cuticular membrane but this does not always stop the penetration of a pathogen. In this instance a zoospore of *Pseudoperonospora humuli* is successfully penetrating just such a young stomata. (Scanning electron micrograph D. J. Royle & G. G. Thomas, Wye College, University of London.)

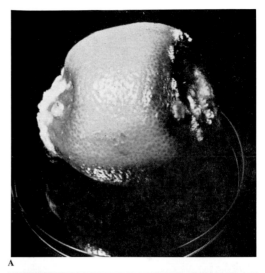

A

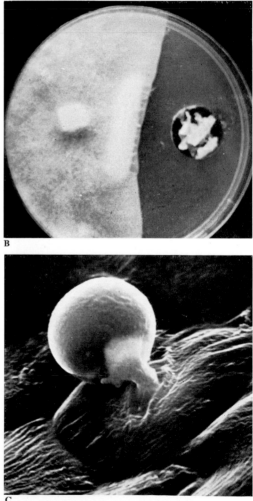

B

C

D

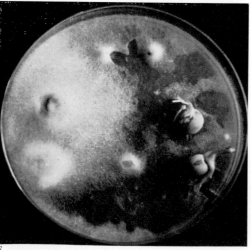

E

F

D Induced chemical defences are illustrated in this experiment where a suspension of *Penicillium* spores was applied to the inner surface of broad bean pods in which the fungus is not pathogenic. The bean tissue responds by producing a whole range of chemicals including brown melanoid pigments. That these are produced in response to the spores is shown by the lack of host reaction to the water applied in the same way (left).

E The hypersensitive reaction is demonstrated when french bean leaves were floated on a suspension of *Botrytis faba* spores, which are not pathogenic to french bean. Microscopically small lesions were produced at the sites of attempted infection. The host reacted violently, killing the pathogen and leaving a small area of dead brown cells surrounded by a ring of dead or moribund cells containing no chlorophyll. (A leaf vein containing little chlorophyll which is not associated with the host–parasite interaction is seen running from top to bottom of the photograph.

F Chemicals can be applied to protect plants against infection. In this instance Captan has been applied to pea seeds (right) and has been successful in protecting them against *Pythium* growing out from the inoculum block (lower centre). The two pea seeds (left) and the two cabbage seeds (top) that were not treated with this dressing have stimulated the growth of the pathogen and have been killed by it.

Plate 10 Host-parasite interaction

In this example a flowering plant parasite,
broomrape (*Orobanche crenata*) is considered.

A The parasite is attached to the roots of a
broad bean plant from which it is drawing
nourishment. Note the large storage organ at the
base of the parasite's flower spike. From this
flower spike the broomrape produces vast
quantities of very tiny seeds.
B These seeds only germinate in the presence
of a suitable host due to their requirement for the
exudate from the host root to trigger germination.
In this instance germination has been triggered
in vitro with a host root extract. In the absence
of a host the seed will remain dormant in the
soil for many years.

A

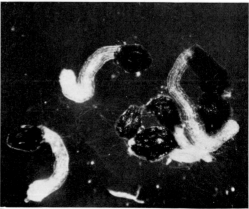

B

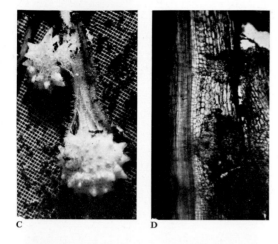

C

D

C   The germination pegs penetrate the host roots where the parasite's cells proliferate forming a tiny tuber that can split open the host root as it grows.

D   As the tuber increases in size it may contact other host roots into which it pushes haustoria to absorb nutrients. The photograph is of a longitudinal section of such a fusion region.

E   Primary and secondary fusion regions differentiate to produce conductive elements for the transport of the large amounts of nutrients required by the parasite. The photograph is a section of this region, bean tissue containing tracheids on the left and broomrape tissue containing tracheids on the right.

F   The transport of carbohydrates from the host to the parasite is clearly shown in this autoradiograph montage. Radioactive carbon dioxide is converted to sugars in the host leaves during photosynthesis, from where it is translocated down to the roots. The parasite, however, rapidly extracts these sugars from the roots and transports them to its own developing flower spike, depleting the host roots of nutrients and ultimately killing the host. (The presence of radioactivity is indicated by the blackening of the film, the degree of blackening being roughly proportional to the amount of radioactive material present).

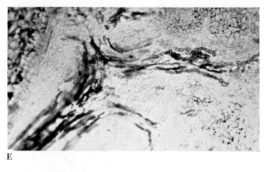

E

F

canadensis contains two alkaloids, sanguinarine and chelerythrine, also at concentrations higher than that needed to stop the growth of the fungus in culture.

Pathogens vary in their sensitivity to alkaloids, *P. omnivorum* being very sensitive whilst *Verticillium* and some of the other vascular wilt fungi are relatively unaffected by them. Disease resistance due to alkaloids is probably the exception rather than the rule. Many plants contain little or no alkaloid and are resistant to a wide range of pathogens, whilst species that do contain alkaloids often show variations in disease resistance quite unrelated to the alkaloid content of the tissue. It has been found, for instance, that although alkaloids accumulate around wounds in a potato tuber, there is no correlation between the rate or amount of accumulation by different varieties and their resistance to infection by a wide range of pathogens following wounding.

Glycosides are compounds with one subunit (aglycone) attached via an oxygen bridge to a sugar molecule. Many compounds in this category produced by plants have anti-microbial activity but as with alkaloids this is thought to be of importance in only a few host-parasite combinations. It is nearly always the aglycone subunit that has anti-microbial activity, and sometimes it is only active after it has been split from the sugar subunit. It appeared that one such example was black mustard (*Brassica nigra*), one of the few members of the *Brassicae* to be resistant to *Plasmodium brassicae,* and it was suggested that black mustard is resistant to infection because of the presence of large amounts of the glycoside sinigrin. Sinigrin itself is not toxic to *P. brassicae* but hydrolysis of the glycoside by the enzyme myrosine released from damaged cells yields the highly toxic aglycone known as mustard oil. Further work however has shown that highly resistant and fairly susceptible varieties of black mustard contain similar amounts of sinigrin and myrosine. Furthermore, when resistant varieties were grown in the absence of sulphur they synthesized no sinigrin (sulphur being needed for the mustard oil aglycone, allylisothiocyanate), but they were still completely resistant to *P. brassicae.* While this observation appears to have destroyed totally the theory of disease resistance due to glycosides, there do seem to be examples of its importance in some diseases. Some recent work with *Ophiobolus graminis* ('take all'), which attacks the roots and lower stems of cereals, has shown that oats are probably resistant to the disease because of the presence of a glycoside which unhydrolysed was found to be highly toxic to the fungus in culture. There is one strain of *O. graminis,* however, which can successfully attack oats, *O. graminis* var *avenae.* In culture it too is inhibited by the pure glycoside from oats but not by the crude extract. So clearly there are compounds in the oat plant which enable this one strain to overcome the inhibition caused by the glycoside, something the other strains are unable to do.

It is not surprising that attempts have been made to discover the rôle, if any, in disease resistance of compounds so central in plant metabolism as amino acids. In some instances there does appear to be a relationship between amino acid content and susceptibility of a plant to infection. The normal range of variation in amino acid concentration has no apparent effect on susceptibility, but when the amino acid content is induced to rise substantially above the normal level some plants show an increase in susceptibility whilst others show a decrease. For example, when cucumber seedlings are placed in amino acid solutions prior to inoculation with *Cladosporium cucumerinum* certain amino acids cause a marked reduction in susceptibility whilst the others have no effect. Most of the amino acids that decrease the susceptibility of cucumber to *C. cucumerinum* do not occur in nature: $\overset{+}{\text{D}}$-serine, $\overset{+}{\text{D}}$-threonine and $\overset{+}{\text{L}}$-threo-β-phenylserine reduced infection whilst the naturally occurring isomers were inactive. Inhibition of fungal growth was not due simply to the unnatural isomer upsetting protein synthesis in the fungus because the effect was far more marked than the slight toxicity of these amino acids to the fungus in culture. No fungitoxic substances could be extracted from the seedlings, but the amino acids may have altered the metabolism of the cucumber seedlings to make them more resistant to infection although the mechanism is not at all clear.

It is not only unnatural isomers of amino acids that alter the host-parasite relationship. When wheat seedlings susceptible to *Puccinia graminis* are infiltrated with thiosemicarbazide, semicarbazide, or fluoracetic acid, all treatments causing a build-up of amino acids, the seedlings become resistant to the fungus. It has been suggested that this is due to the increase in amino acid concentrations but equally resistance might arise from some other effect of treatment with these metabolically very active compounds. More direct evidence for the importance of amino acids in some infections is that cut apple shoots treated with $\overset{+}{\text{D}}$-phenylalanine, $\overset{+}{\text{D}}$-alanine and $\overset{+}{\text{DL}}$-α-aminobutyric acid had smaller lesions as the result of infection by *Venturia inaequalis* than untreated shoots. It is interesting to note that the most active amino acid in this instance, $\overset{+}{\text{D}}$-phenylalanine, had no effect on cucumber seedlings.

In contrast to these findings it has been observed that there is an inverse correlation between susceptibility to *Fusarium oxysporum* and the amino acid content of lupin seedlings. Sap from susceptible plants contains more leucine, aspartic acid, alanine and asparagine than resistant plants, and it also contained glutamic acid which was not detectable at all in resistant plants. Infection results in a decrease in the amino acid content of the sap of susceptible plants and this may reflect the utilization of the amino acid by the fungus; resistant plants owe their resistance to the fact that the fungus grows poorly in the absence of these amino acids.

Suggestions and hypotheses based on such observations are clearly speculative and it is only when these pieces of information start to fit together to form a coherent whole that a theory can be put forward with any degree of confidence. Unfortunately physiological plant pathology is in its infancy compared to many sciences so that as yet only a few of the pieces fit together and many of the hypotheses are put forward merely as an anvil on which to hammer out new experimental approaches.

Proteins are even more central in the metabolism of a plant than amino acids. As enzymes they are involved in practically every synthetic and degradative process, including the production of defence mechanisms and the breakdown of microbial toxins. So far there is not a great deal of evidence for the direct involvement of proteins in the host-parasite interaction. It has been suggested that highly obligate host-specific pathogens are only able to attack a plant which contains specific proteins for which the pathogen has a requirement, and which it is unable to synthesize for itself. Serological analysis has been used to investigate the proteins of host plants and in some cases it has been found that susceptible varieties contained a particular protein absent in the resistant varieties. These and similar findings have led to the 'gene for gene' hypothesis. This suggests that the protein conferring susceptibility or resistance is coded for by a particular gene in the host and the requirement for this protein is coded for by a particular gene in the parasite, leading to the very high degree of specificity observed in some diseases. A slight modification of this hypothesis is that the host contains a gene coding for a particular protein while the pathogen lacks its own functional gene coding for this protein which is nevertheless essential for its own metabolism. Although in each instance the word 'protein' is used, this does not exclude the possibility that in fact a protein subunit is involved in the interaction.

A considerable amount of work has been done on the level of carbohydrate in the host tissue and its resistance to infection. It has proved possible, on a purely empirical basis, to divide the majority of plant disease into one of two groups, high sugar diseases and low sugar diseases. A high sugar disease (eg rusts and powdery mildews) is one in which the plant is most susceptible when its tissues contain a high concentration of sugars while a low sugar disease (eg dutch elm disease, *Fusarium* wilts and early blight of tomato) is one in which the host is most susceptible when its tissues contain low concentrations of sugars. This is not only based on observations of sugar concentration at times of high and low susceptibility, but on experimental situations where the carbohydrate level was manipulated, either by altering the photosynthesis of the plant or by directly feeding sugars to the tissues of the plant. In practically all cases there was the expected change in susceptibility. Boron deficiency, which increases the accumulation of sugars in the leaf, also has the predicted effect on disease resistance. Similarly, the removal of carbohydrate sources, such as leaves, and carbohydrate sinks such as fruits, will alter the carbohydrate level in the host and again the resultant changes in susceptibility are the expected ones.

The theory is that high levels of available carbohydrate in the host tissue may in certain diseases increase both the vigour of the pathogen and its ability to infect the host (high sugar diseases). In other diseases the effect is reversed: an abundance of carbohydrate in the host tissue enables it to respond quickly and efficiently to an attempted invasion by a pathogen, and lower levels of carbohydrate reduce the host's ability to resist infection (low sugar diseases).

Unfortunately, as with most of the hypotheses attempting to explain host-parasite interaction, there are indications that this one is a little too simple. Treatment with plant hormones, for example, will alter the sugar balance of the host plant but does not always have the expected effect on susceptibility. Nevertheless, there is no doubt that the hypothesis is useful and that many plant diseases are influenced by the carbohydrate level of the host.

Although there is relatively little work reported on the importance of hydrogen ion concentration (pH), it is probably of considerable importance in some diseases.

As bacteria are generally more sensitive to pH in culture than are fungi it seems likely that pH is more important in affecting the susceptibility of plants to bacterial infections than to fungal infections. The pH of many plant tissues is, in fact, low enough to inhibit the growth of a large number of bacteria and a few fungi. The pH of ripe tomatoes (4·0-4·6) is low enough to inhibit the growth of *Xanthamonas vesicatoria,* and is thought to be the reason for the fruit being resistant when ripe but susceptible when unripe, at which time the pH is somewhat higher (over 5·0) and is not inhibitory. The resistance of tomatoes to *Fusarium culmorum* and *F. oxysporum* is also thought to be due to pH resistance increasing as the fruit ripens. The resistance of water melon to *F. niveans* may be due to the high concentration of acetic acid in the host tissue.

It is not just the pH of the tissue that must be considered in this context but also the ability of the tissue to maintain this pH, i.e. its buffering capacity. *Colletotrichum hibisci* grows well in culture only at around neutrality but it is capable of germinating over a wide pH range (3·0-8·5). The fungus, although not growing vigorously, quickly increases the pH of an acid medium to neutrality or even slight alkalinity at which time rapid growth ensues. *Hibiscus cannabis* is attacked by *C. hibisci*; this is not surprising in the seedling which has a uniformly slightly alkaline pH, but it is surprising that the mature plant is attacked as the shoot has a pH as low as 4·0 and would on this account be expected to reduce the growth of the pathogen. However, the ability of the fungus to alter the pH of the medium in which it is growing enables it to colonize the mature *Hibiscus* shoot efficiently.

When considering pH one must also bear in mind the variations that can occur between different organs of the plant, in different parts of a plant cell and at different times of day, as well as those taking place during growth and maturation of the host.

5.4   Infection-induced biochemical defences

The term phytoalexins was coined by Müller in 1956 to describe fungitoxic or bacteriotoxic compounds produced by a plant in response to injury or infection. The name was applied initially to compounds produced solely in response to injury or infection and did not include compounds normally present in a healthy plant and synthesized in larger quantities following damage. However one or two compounds known for some time as phytoalexins are now thought to have been detected in exceedingly tiny amounts in healthy tissues. Until recently it has been thought that all phytoalexins were produced only around the infected or damaged area and were not translocated away from this site, but there is now some slight evidence that some of the new phytoalexins being discovered are in fact translocated to give generalized protection in response to a localized infection.

Phytoalexins do not seem to be related specifically to the pathogen in the same way as animal antibodies. This means that stimulation by one micro-organism resulting in the production of phytoalexins will tend to increase the resistance of the plant to infection by other micro-organisms.

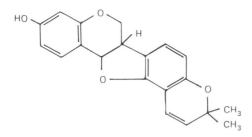

A PHASEOLIN (from beans)

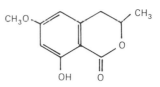

B ISOCOUMARIN (from carrots)

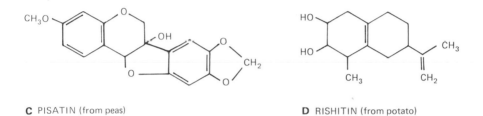

C PISATIN (from peas)

D RISHITIN (from potato)

Figure 5.9 Examples of four phytoalexins. **A** Phaseolin (from beans) **B** Isocoumarin (from carrots) **C** Pisatin (from peas) **D** Rishitin (from potatoes).

The production of pisatin and phaseolin can readily be demonstrated by applying a droplet of spore suspension to the inside of a pea or bean pod. The tissues of the pod respond by producing fungistatic compounds which are not produced when a drop of water alone is applied. The reason for choosing a pea or bean pod as the experimental material is that a sterile surface can easily be exposed without it suffering damage. This is important because mechanical damage can itself sometimes stimulate the production of fungistatic compounds. There is now some evidence that surfaces not normally exposed to the early stages of infection, such as the inner surface of a pod, do not respond in exactly the same way as the surfaces of stems and leaves, so that there is a trend towards using more normally infected surfaces for research work. The problem is that plants grown under normal conditions are exposed to an enormous number of micro-organisms which are not able to overcome the plant's defences and infect the plant, but which do stimulate the production of phytoalexins. This makes it difficult to distinguish between preformed biochemical defences and those produced in response to a potential pathogen, and it also makes it difficult to distinguish between phytoalexins produced in response to the experimental treatment and those produced in response to the normal micro-flora. The answer is to grow plants under sterile conditions, in itself an abnormal situation, and under these conditions it can be difficult to maintain a good growth environment. As light, temperature and humidity also affect the host-parasite interaction it is necessary to use equipment providing a sterile, controlled environment which is costly and time consuming to use but essential for many experiments.

The production of phytoalexins is much lower in plants attacked by a highly specialized obligate parasite than in plants attacked by less specialized pathogens. Interestingly the greatest phytoalexin production is stimulated by organisms not able to establish an infection at all on that particular host. It is perhaps because they stimulate such a violent reaction from the host that their attack is unsuccessful. The same pattern is true of virulent and avirulent strains of the same pathogen, the avirulent strain stimulating the greatest phytoalexin production. The virulent strain stimulates a lower rate of phytoalexin synthesis, with a reduced final yield, so that it is well established in the host before appreciable amounts of phytoalexins are produced.

Although mechanical damage can sometimes stimulate the production of anti-microbials, it is not just the mechanical damage caused by the penetration of a pathogen that triggers production. The metabolic by-products of the pathogen seem to be particularly important in stimulating phytoalexin production by the host cells. The application of filter-sterilized culture medium, in which the micro-organism has grown, will stimulate plant cells to produce phytoalexins. Quite a wide range of other compounds such as the ions of copper, silver, and mercury can also stimulate the production of phytoalexins.

The amount of phytoalexin produced depends not only on the micro-organism attempting to invade the host but also on the amount of inoculum present. There is almost a linear relationship between pisatin production and the number of conidia in the infection drop often over a range as wide as $3 \times 10^3 - 3 \times 10^5$ spores/ml.

Pathogens able to infect a particular host, as well as stimulating less phytoalexin production than non-pathogens, are also less sensitive to its toxic effects. The ED50 (i.e. the minimum dose required to affect 50 per cent of the population) for *Sclerotinia fructicola* (non-pathogenic on french beans) is about 3 μg/ml of phaseolin, whereas the ED50 for *Colletotrichum lindemuthianum* (causing anthracnose of beans) is over 50 μg/ml. Pathogenic strains might be able to degrade or detoxify the phytoalexin. Two pea pathogens *Fusarium solani* f. *pisi* and *Ascochyta pisi* completely degraded pisatin in three days when it was added to the liquid culture in which they were growing; *Sclerotinia fruiticola* (non-pathogenic to peas), on the other hand, only degraded about half the pisatin in that period.

Phytoalexins can be summarized as a group of compounds which are:
(i)   Fungistatic and bacteriostatic, and are active at very low concentrations;
(ii) Produced by the host plant in response to infection or in response to the metabolic by-products of micro-organisms and to a few other stimuli;
(iii) Absent from healthy cells or only occur in the most minute quantities;
(iv) Usually remain close to their site of production;
(v)   Produced in quantities proportional to the size of the inoculum;
(vi) Produced in larger quantities in response to weak pathogens and non-pathogens than to virulent pathogens;

(vii)   Produced relatively quickly by the cells, usually within twelve to fourteen hours, reaching a peak around twenty-four hours after inoculation;

(viii) Host specific, rather than pathogen specific.

Most compounds toxic to micro-organisms which are present in healthy plant tissue are produced in increased quantities following infection. The importance of polyphenols as a biochemical defence mechanism has already been mentioned and these are among the compounds whose production is stimulated by infection. It has been shown that when sweet potatoes are infected with *Ceratocystis fimbriata* the concentration of a number of fungitoxic substances, including the polyphenols chlorogenic acid and isochlorogenic acid, increases just ahead of the invading fungus. Similarly potato tissues synthesize caffeic acid and chlorogenic acid in response to *Helminthosporium carbonum* infection but not in sufficient quantities to inhibit the pathogen totally. It is thought that there is a synergistic action between these compounds and other compounds also only slightly toxic to the fungus.

There is an interesting situation in the infection of resistant and susceptible potatoes by *Phytophthora infestans*. Both resistant and susceptible tubers contain caffeic and chlorogenic acids when healthy, the resistant tubers containing 1·2 mg/g of chlorogenic acid whereas the susceptible tubers contain only 0·3 mg/g. Following infection, however, the chlorogenic acid content of the susceptible tubers rises to about 1·2 mg/g and the concentration in the resistant tubers falls to about 0·5 mg/g. This phenomenon, coupled with the fact that *P. infestans* is highly tolerant of chlorogenic acid in culture, seems to indicate that chlorogenic acid plays little part in *Phytophthora* blight resistance. It is nevertheless possible that it is involved indirectly as an intermediary in a resistance mechanism. Some polyphenols may act both directly and indirectly, and an example of this behaviour is seen with caffeic acid which is both toxic to many fungi and is an intermediary in the pathway to lignin synthesis:

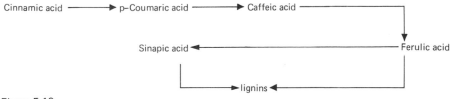

Figure 5.10

As well as polyphenols there are a considerable number of other fungitoxic and bacteriotoxic compounds produced in greater quantities by plant cells as a result of infection. Some of these are phenolic, with only one OH group, such as isocoumarin which is produced in large quantities by some tissues following infection; for instance, carrot infected by *Ceratocystis fimbriata* produces 0·6 mg/g of isocoumarin. As a concentration of 0·2 mg/g will inhibit 95 per cent of the growth of the fungus in pure culture it seems likely that this is a major defence mechanism of carrot against attack by *Ceratocystis*.

There are cases where the host tissues synthesize phenyl glycosides which are not toxic to the pathogen until enzymes liberated by the pathogen hydrolyse it and release the toxic phenol. *Venturia pirina,* which causes pear rot, produces a β-glycosidase which catalyses the hydrolysis of arbutin to liberate hydroquinone in toxic quantities.

The increased production of toxic compounds is frequently the result of the increased synthesis of particular enzymes. For instance, the production of the fungitoxic brown melanoid pigments by the oxidation of polyphenols is in some instances merely due to mechanical damage allowing the polyphenols, poly-phenoloxidase and oxygen to react in a way that would not be possible in an undamaged cell. But this is not always the explanation because frequently there is a measurable increase in the amount of polyphenoloxidase present in the tissue, presumably synthesized in response to invasion. There appears to be a clear relation-ship between the increased synthesis of this enzyme and disease resistance of some plants. Resistant potatoes infected with *Phytophthora infestans* have high levels of polyphenoloxidase in the four days following infection and its production extends for a considerable distance from the site of inoculation. Treated susceptible potatoes, on the other hand, show little increase in polyphenoloxidase activity. Following inoculation with *Fusarium oxysporum* f. *pisi,* resistant peas also have levels of this enzyme which are about three times as high as those found in similarly treated susceptible plants. Resistant and susceptible tomatoes show the same pattern of increased polyphenoloxidase when inoculated with *Fusarium oxysporum f. lycopersici.* The resistance of many plants to a considerable number of diseases decreases if they are treated with a reducing agent prior to inoculation, presumably by counteracting the effect of the oxidase enzymes. More specifically, 4-chlorore-sorcinol decreases the resistance of *Malus atrosanguinea* to *Venturia inaequalis* by inhibiting polyphenoloxidases.

Peroxidase is another enzyme implicated in disease resistance. Although its rôle is not known the correlation between disease resistance and high peroxidase activity is sufficiently good for it to be used as a preliminary screening for potatoes bred for resistance to *Phytophthora* blight. Potatoes are not the only plants to show a correlation between resistance and peroxidase levels since it also occurs in a wide range of plants, such as cabbage in relation to *Botrytis cinerea* infection, in which resistant plants have about twice the peroxidase activity possessed by susceptible plants.

Catalase is also thought to be involved in the same sort of way. Susceptible wheat produces only a slight increase in catalase following infection by *Puccinia graminis* var. *tritici* whereas there is a sharp increase when resistant varieties are exposed to the pathogen. These same resistant varieties also show increased ascorbic acid oxidase and to a lesser extent cytochrome oxidase levels after infection.

As we have noted there is a rise in activity of a number of oxidative enzymes following infection, particularly in resistant plants, and this undoubtedly accounts at least in part for the increased respiration of many infected plants. Although they are possibly the most important factors, it is not only oxidative enzymes that are involved in the resistance mechanism and are stimulated by infection; every one of the biochemical and histological defences mentioned are produced only as a result of a whole chain of enzymatic changes. Every link of this chain must therefore be strengthened to cope with the increased synthesis required to resist an attack by a pathogenic organism.

5.5 Cytological defence reactions

A defence reaction frequently referred to as plasmatic is one involving the digestion of the invading organism by the cytoplasm of the host cells. This process frequently results in the destruction of the host cell itself but if the pathogen is successfully destroyed it is a very effective way of stopping further infection. The cell constituents of both host and pathogen are absorbed by surrounding cells and are therefore not lost by the host plant. Examples of this type of reaction are found in the endophytic nitrogen-fixing bacteria, *Bacterium radicicola,* and in the mycorrhizal association of fungi with plant roots. Because the nutrients synthesized by the pathogen are of benefit in some cases to the host, following digestion of the hypha by the host cell, such a relationship can be to the advantage of both host and

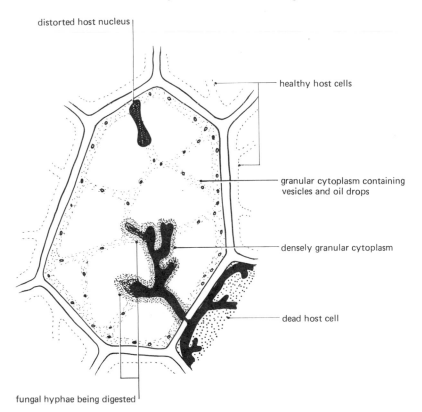

Figure 5.11 Digestion of a fungal hypha by a host cell. This may or may not be successful in halting the growth of the pathogen.

micro-organism and is termed symbiotic. The establishment of satisfactory relation-ship depends on the virulence of the pathogen and the resistance of the host. If the pathogen is too virulent a general invasion of the host occurs whilst if the host is too resistant no association develops.

The delicate balance necessary for a symbiotic relationship is comparatively uncommon and even the host resistance resulting from digestion of the invading parasite occurs only in a limited number of host-parasite combinations. More frequently the host cell reacts violently to invasion, killing both itself and the pathogen before digestion can take place. Such a reaction is called *necrotic,* because of the rapid death of the cells involved, but an even more violent reaction is usually produced by the attempted invasion by a non-pathogenic organism, and this is called *hypersensitive necrosis.*

The necrotic defence reaction is, of course, very effective against highly specialized obligate fungi, bacteria, and especially viruses, by depriving them of their very specialized requirements only available in a living cell. One might suppose that the necrotic reaction would make the invasion of a plant easier for a pathogen which was a facultative saprophyte by providing readily available nutrients to increase its inoculum potential. This however is not the case because as the host cells die substances such as brown oxidized polyphenols and phytoalexins are produced which are strongly inhibitory to the invading micro-organism. Hyper-sensitivity was first noticed in 1902 but its importance has only recently been realized, and it is now considered to be one of the most important of a plant's defence mechanisms. It is directly linked with many of the biochemical defences induced by infection. Phytoalexins seem to be produced almost exclusively by cells undergoing the hypersensitive necrotic reaction and the faster the hypersensitive reaction the greater the production of phytoalexins; or perhaps it is the other way round, and the greater the production of phytoalexins the more rapidly the host cell is killed.

The more resistant a plant is to a particular pathogen the faster the necrotic reaction tends to be. The almost successful invasion of a plant by a pathogen usually results in the necrosis of an appreciable number of cells before the invasion is halted. The result is a necrotic lesion clearly visible to the naked eye and which is frequently used in bio-assay experiments. Tobacco necrosis virus produces just such lesions on french beans. When a plant is highly resistant, as most plants are to the majority of diseases, then the reaction is highly hypersensitive and there is very rapid necrosis of the cell penetrated usually leading to the death of the pathogen within that single cell. Only one cell dies, something which is not easily seen by the naked eye.

The rapid death of plant cells following infection is related to a stimulation of many metabolic pathways, particularly the respiratory pathways. The hypersensi-tive reaction produces a rapid increase in respiration following infection which declines back to normal once the invasion of the host tissue has been halted. The raised respiration is qualitatively different from normal respiration, responding differently to inhibitors of respiratory enzymes and to agents which uncouple oxidative phosphorylation. The raised respiration is not just the result of uncontrolled oxidation of available substrates as the regulatory mechanisms of the dying cell cease to function, because the use of the respiratory inhibitors can alter the tissue's ability to resist invasion. If the increase in respiration that normally follows the inoculation of resistant potatoes with *Phytophthora infestans* is controlled by respiratory inhibitors the resistance is appreciably lowered. Tubers that normally show a violently hypersensitive reaction and stop any infection within one or two cells, become more susceptible following treatment with respiratory inhibitors, the hypersensitive necrotic reaction is much weaker and a considerable amount of tissue is invaded before the infection is halted. Clearly the raised metabolic rate reflected by the increased respiration of hypersensitive necrosis is a vital part of the defence reaction. Perhaps it is connected with the synthesis of the anti-microbial compounds already noted.

Many of the processes involved in a plant's defence against microbial attack are intimately interlinked, which is perhaps not surprising when one remembers that they all depend on the biochemical processes going on within plant cells which are in turn controlled by the genetic information passed on from previous generations. It is this handing on of genetic information from one generation to another that leads to the situation where most plants are resistant to the majority of micro-organisms, since plants highly susceptible to a large number of microbial pathogens would surely soon become extinct.

Chapter 6
Disturbance of the host metabolism (The causes of symptoms and death)

The massive destruction of host tissue by unspecialized parasites growing through cells, rupturing their membranes and destroying their osmotic integrity, is an obvious cause of the visible symptoms in plants. As parasites become more specialized the damage they do to the host is less traumatic and the infected tissues may survive for a considerable period. In most instances even specialized parasites cause a disturbance to the host metabolism or trigger a defence reaction by the host which may lead to the death of a small amount of tissue or to the ultimate death of the host.

The following sections in this chapter deal with some of the disturbances to the host metabolisms that result in the symptoms of which we are so aware in the diseased plant.

6.1 Pathogen-produced toxins

There is a wide range of toxins produced by plant pathogens and they vary in the degree with which they are implicated in the disease syndrome. Their effects range from those which produce all the disease symptoms and have the same host specificity as the pathogen, to those compounds which, although phytotoxic, do not induce the same symptoms in the host and do not have the same host specificity as the pathogen producing them. It is this first category that is thought to be of relevance and therefore has been the most extensively studied, but there are many toxins causing some, but not all of the disease symptoms, and these together with other host-parasite interactions may be responsible for the observed symptoms.

The study of toxins is complicated by the fact that many micro-organisms liberate different compounds under different growth conditions. The fact that a pathogen does not produce a toxin in culture does not mean that it will not produce it in the host, and conversely the toxins produced in culture may not be produced in the host. Toxins have been isolated from diseased plant material that were produced neither by the pathogen in pure culture nor by the healthy host. Either the toxin is produced by the host in response to the pathogen or it is produced by the pathogen under the special conditions occurring in the host. It is extremely difficult to distinguish between these two possibilities unless the conditions of pure culture of the pathogen can be adjusted so as to induce production of the toxin.

Victorin is a good example of a pathogen-produced toxin. It is produced by the fungus *Helminthosporium victoriae* which causes the victoria blight of oats. The amount of toxin produced is related to the virulence of the infecting strain, highly virulent strains of the fungus producing large amounts of the toxin when grown in culture and avirulent strains producing relatively little. Victorin is thought to be a polypeptide linked to a tricyclic secondary amine but despite our knowledge of its chemistry the mechanism of its toxicity is not at all clear. It causes a marked increase in the respiration in all the tissues of susceptible oat plants but has no effect on the respiration of resistant plants. The increase in respiration of suscept-ible oat tissue is proportional to the concentration of the toxin but the latter appears to have no effect on the respiration of mitochondria isolated from suscept-ible plants, nor does it affect the succinic oxidase activity of these mitochondria. It

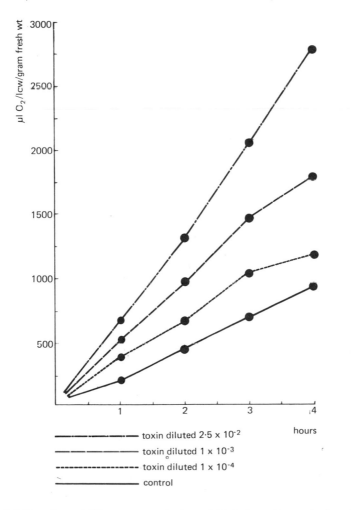

Figure 6.1 The effect of different concentrations of Victorin on the respiration of oat tissue. (after Krupa, 1958)

does however stimulate their oxidation of ascorbic acid. The increased respiration may be an indirect effect of the toxin. This idea is borne out by the increase in permeability of host cells after treating them even with very low concentrations of Victorin, lower than those causing changes in respiration. The increased respiration may in fact be the result of this increased permeability of cell membranes. Interestingly the toxin renders the oat tissues insensitive to dinitrophenol (DNP), an uncoupler of oxidative phosphorylation. It is therefore possible that a rate-limiting step is already the controlling factor in the respiration of Victorin-treated oat tissue and that a fall in adenosine triphosphate (ATP) levels as a result of the uncoupling of oxidative phosphorylation by DNP cannot cause any increase in respiration. Other pathogen-produced toxins besides Victorin also frequently cause this increase in respiration and permeability.

As previously mentioned, not all toxins produce the entire range of disease symptoms, but there are examples of the entire range of symptoms being produced by a combination of toxins produced by a pathogen. For example, the vascular wilt disease caused by *Fusarium oxysporum* results in wilting, browning and leaf necrosis, as well as vascular blockage of quite a wide range of host species. Compounds produced by fungus in pure culture will induce all these symptoms when a cutting of the plant is placed in a sterile culture filtrate. The wilting is caused by phytonivein, the leaf necrosis is caused by fusaric acid and pectic enzymes appear to be the cause of vascular blockage. These compounds singly or together do not show the same host specificity as the pathogen, i.e. will cause symptoms in plant varieties resistant to *Fusarium oxysporum,* so in this instance although the full range of symptoms can be produced by the toxin, the host-parasite relationship clearly involves more than just toxin production.

Another toxin that has been extensively studied is produced by *Pseudomonas tabaci,* the organism causing wildfire disease of tobacco. The toxin causes chlorosis of the leaf tissue and in an infection by *P. tabaci* the toxin diffuses out in advance of the water-soaked area in which the bacterium multiplies producing a chlorotic ring surrounding a water-soaked necrotic centre. The toxin is again non-specific, causing symptoms in resistant as well as susceptible species, and in fact, it will even cause chlorosis of unicellular green algae, and this is used as the basis of a bio-assay technique with the alga *Chlorella.* There is an approximately linear relationship between the growth of the alga and the concentration of wildfire toxin, but the inhibitory effects of the toxin can largely be overcome by the addition of methionine to the culture medium. Experimental results have clearly shown the competitive inhibition of the toxin by methionine and it was therefore reasoned that the toxin either interfered with methionine production or with its utilization, affecting protein synthesis which in turn ultimately resulted in the chlorotic symptoms caused by the toxin. Circumstantial evidence for a connection between protein synthesis and chlorosis is provided by the plant hormone kinetin, which has been shown to stimulate protein synthesis in leaf tissue and to delay chlorosis and senescence in excised leaf discs.

Elucidation of the structure of wildfire toxin further explained its competition with L̄-methionine. They are structurally related and might well compete for sites

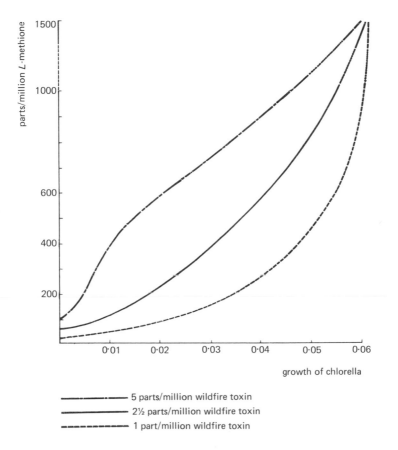

Figure 6.2 To show the competitive inhibition of wildfire toxin by L-methionine using the growth of *Chlorella* as a bio-assay. (after Braun, 1955)

on enzymes involved in methionine utilization. Methionine sulphoximine also competes with methionine in protein synthesis, and it too causes chlorosis when applied to plant tissues.

Although Victorin is a polypeptide with a secondary amine side chain, and the wildfire toxin is a modified dipeptide, not all toxins are based on amino acid polymers. Pathogen-produced toxins do not all fall into a single group either structurally or functionally, and they show a wide range of structures and modes of action.

A number of the toxins produced by bacteria are polysaccharides, or polysaccharide derivatives, which are often slimy or mucilaginous. It has been established in some instances that there is a direct relationship between pathogenicity of a bacterium and its production of polysaccharide slime. Four forms of *Xanthomonas phaseoli,* for example produce different amounts of slime and their pathogenicity is proportional to these amounts. This does not prove that the slime

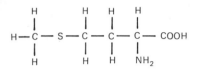

A *L*-METHIONINE

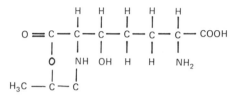

B WILDFIRE TOXIN

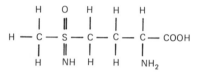

C   METHIONINE SULPHOXIMINE

Figure 6.3 Showing structural relationship of **A** L-methionine, **B** wildfire toxin and **C** methionine sulphoximine.

is the cause of symptoms in the infected bean plant, since they could be due to some other factor that happens to be proportional to the production of slime, or the correlation might be purely chance and the symptoms quite unrelated to slime production. The same correlation with pathogenicity has however been found in isolates of different species producing varying amounts of slime. *Pseudomonas solanacearum* is found in three strains in which pathogenicity is related to slime production. Thus the evidence for a direct relationship between pathogenicity and the production of slime builds up but it was not until the polysaccharides were purified and found themselves to induce disease symptoms when injected into the host that the relationship was in any way certain. Polysaccharide slime was purified from liquid cultures of *Corynebacterium insidiosum.* The purified slime at concentrations as low as 20 µg/ml, would induce wilting symptoms in lucerne cuttings ten to fifteen minutes after application to the cut end of the shoot.

Similar results were found with the polysaccharide isolated from *Agrobacterium tumefaciens* when supplied to tomato cuttings.

The slime acts by blocking the vascular system of the host, but the site of blockage is unknown. Blockage occurs fairly quickly since removal of the lower few centimetres of a cutting placed in a slime solution enables the cutting to recover turgor when it is then replaced in water. If the lower few centimetres is not removed there is no recovery. This hypothesis does not explain some of the symptoms observed, such as the local lesions produced by *Xanthomonas phaseoli.*

Although the mode of action of many of these polysaccharide slimes may be similar they vary greatly in their chemical composition and size. The slime produced by *Agrobacterium tumefaciens,* for instance, consists of twenty-two glucose units and has a molecular weight of about 3600, whilst that produced by *Xanthomonas phaseoli* is composed of glucose, mannose and glucuronic acid and has a total molecular weight of around 19 000 000.

6.2 Interferance with water economy

Insufficient water reaching the leaves of a plant is indicated by signs of wilting and flaccidity in the tissue. It makes no difference to the symptoms whether the insufficiency is due to a shortage of water in the soil, the inability of the roots to take it up, the inability of the vascular system to transport adequate amounts of water to the leaves, or to a disturbance of leaf metabolism causing excessive water loss which cannot be matched by the supply system.

Wilting occurs in two stages, reversible and irreversible. Reversible wilting means that temporarily the water loss by the leaf has exceeded the supply. It happens frequently in healthy plants during the heat of a summer's day and is rapidly reversed as the air temperature falls and the humidity rises. It does no harm to the plant and is often part of the normal pattern of water conservation because when a leaf is wilted the guard cells become flaccid and close the stomatal opening, and the projected area on which the sunlight is falling is reduced as the leaves droop. If, however, the wilting is prolonged or very severe, from whatever cause, permanent damage is done to the plant and it may loose the ability to recover its turgor. Wilting has then become irreversible. Reversible wilting is frequently the first sign of water stress in an infected plant and they frequently recover temporarily from this stage when the environmental conditions become more favourable either by reducing the water loss or by increasing the water availability. Reversible wilting is usually followed by irreversible wilting and death of the affected leaves. Death of the larger leaves usually occurs first followed by the younger leaves and finally the growing points. This sequence perhaps reflects the water loss from the various parts of the infected plant.

In 'damping', 'foot rots' and 'take all' diseases there is a generalized destruction of root and basal stem tissues by unspecialized parasites. The uptake and transport of water is thus inhibited but no wilting occurs before obvious damage has been done to the host. In wilt diseases, such as those caused by *Fusarium* and *Verticillium,* and by some viruses, wilting is one of the first symptoms and is obvious long before any gross damage to the plants can be observed. This may be due to the pathogen reducing the ability of the roots to take up water, to the inability of the vascular system to transport it or possibly even to uncontrolled water loss from the leaves. It has been shown that when pepper seedlings were infected with tobacco etch virus wilting was accompanied by a loss of electrolytes from the root which started some twenty four hours before wilting was visible. Pepper varieties which, while susceptible to the virus, do not show these particular symptoms of wilt do not lose electrolytes from the roots. The loss of electrolytes clearly shows an altered permeability of root membranes and this could easily be the cause of reduced water uptake by the roots giving rise to water stress in the shoots.

The other possible causes of wilting do in fact occur. In vascular wilt diseases, as the name implies, the effects of the pathogen are largely in the vascular system and it is there that the majority of the mycelia or bacteria are found, but interference with water transport is probably not caused simply by the mass of the pathogen blocking the vessels of the vascular system. Even the most prolific vascular wilt fungi do not produce enough mycelia to block a very large percentage of the vessels or to cause a very significant increase in resistance to water flow. Indeed a great many of the vascular wilt fungi, including those that cause very rapid development of symptoms, produce relatively little mycelia. It is generally agreed that in some instances the production of cellulolytic and pectolytic enzymes by the pathogen can result in blockage of the host's vascular system by partially hydrolysing the cellulose and pectin of the host cell walls with the resulting gel blocking the vascular system.

This concept does not however explain all the host-parasite interactions. *Verticillium albo-atrum* produces cellulases and pectinases only in the presence of a suitable substrate; i.e. they are inducible enzyme systems triggered by the presence of the substrate. Induction takes place only in the absence of more easily utilized respiratory substrates such as sucrose, glucose, fructose, maltose, etc. These sugars totally inhibit cellulase and pectinase production so that it would be expected that in a plant containing large amounts of sugars little or no cellulase or pectinase would be produced by *Verticillium*. This is exactly what is found to happen: cellulase and pectinase are produced by the fungus only in the latter stages of the disease, when the plant is dead and the sugars exhausted. The fungus can then utilize the cell wall materials as respiratory substrates for the synthesis of fungal resting structures.

In some instances it is possible that wilting is due to the production by the host of tyloses and gummoses which block the xylem elements of the vascular system. We have considered this process as part of the host defence mechanism against the growth of mycelium in the vascular system, and the transport of conidia, bacteria and toxins in the transpiration stream. The corollary of stopping this movement of the pathogen can be an inadequate water supply to the leaves. This may be a small price to pay if the localization of the infection is successful and regrowth can repair any damage caused by it, but if the defence mechanism is too extreme it may result in the death of the shoot from desiccation.

The production of gelatinous polysaccharides by the pathogen also contributes to vascular disfunction, resulting in wilting. As we saw previously, there is a relationship between the severity of the symptoms caused by a number of pathogens and the amount of polysaccharide slime they produce. Studies on the way that slimes, gums and gels from whatever source cause wilting has shown that in a number of plants they induce two forms of wilting: (i) which affects leaflets but leaves stems and petioles turgid; and (ii) a generalized wilting which particularly affects stems and petioles. In general it seems that water soluble compounds tend to produce the first type of wilting whereas less water soluble compounds tend to produce the second type of wilting.

An experiment with polyethyleneglycols of different molecular weights showed that the lower molecular weight polymers, which had a high water solubility, caused leaflet and leaf edge wilting, whilst the high molecular weight polymers, with a lower water solubility, caused generalized wilting. It is thought that the higher molecular weight polymers caused occlusion of the more major vessels of the stem and petiole whilst the smaller molecular weight polymers passed relatively easily through the major vessels but blocked the tracheids in the fine vascular branches at the edges of leaves and leaflets. A slight modification of this theory suggests that these polymers do not block the vascular elements but cause a marked increase in the viscosity of the xylem sap, thereby reducing its flow rate.

In a number of diseases caused by fungal, bacterial and viral pathogens the stomata close and transpiration is reduced some time before wilt symptoms become visible. *Fusarium*-infected tomatoes transpired only one third as rapidly as healthy plants, the stomata of infected plants remaining closed throughout the day whilst those of healthy plants were fully open for most the daylight hours. In other diseases the rate of water loss is increased. Mildews and rusts cause an increase that is thought to be due to their rupture of the host epidermis and cuticle. In such cases the increased loss is solely through the epidermis and cuticle, and transpiration through the stomata is unaffected so long as there is an adequate water supply. There are occasionally cases where stomata transpiration is increased as a result of infection. In the early stages of infection by *Pseudomonas solanacearum* there is an increase in transpiration by a disturbance to the metabolism of the guard cells causing them to stay open at times when they would normally close.

The primary cause of death in vascular wilt infections may be due to desiccation but it is possible that in many cases it is due to a build up of toxic materials that are usually lost in a gaseous form during transpiration. For instance the build up of ethylene that would occur whilst the stomata were closed would be highly deleterious to the host metabolism. In the absence of photosynthesis the closed stomata would also lead to a build up of carbon dioxide and a reduction in oxygen concentration causing anaerobic respiration, the by-products of which (organic acids or ethanol) might accumulate to lethal levels.

Water not only travels up a plant in the transpiration stream but down the plant during the translocation of sugars in the phloem. This form of water transport is also affected by plant pathogens but the movement of water is of comparatively little importance compared to the disruption of sugar transport.

The rust infection of beans caused by *Uromyces phaseoli* produces abnormal phloem transport in which the nutrients are moved towards the site of infection depleting the surrounding healthy tissues of starch and sugars. There is also evidence that there is a long distance transport of nutrients to supply the increased demand of the infected region. This means the pathogen will adversely affect the growth of the uninfected parts of the plant, reducing its vigour and yield. The infection by the flowering plant parasite *Orobanche* is thought to cause wilting and death of the host by the re-direction of both sugars and water from the host to the parasite. The carbohydrate-starved roots grow poorly and are unable to meet the water requirements of both host and parasite and this results in the death of the host by desiccation. Phenomena such as these are almost invariably associated with highly specialized obligate parasites, less specialized parasites usually having a simpler effect on the host metabolism.

Unspecialized parasites will also on occasion produce symptoms of water stress. The infection of forest tree roots by *Phytophthora cinnamoni* can result in the complete destruction of lateral roots. Growth of new roots to replace those lost often means that the disease is chronic rather than acute and in some instances the trees may not be killed but merely stunted in growth due to water shortage. Paradoxically this disease usually occurs in waterlogged soils where there is an excess of available water, the zoospores apparently being attracted to the tree roots by the ethanol they produce during anaerobic respiration in badly aerated soils.

Many other root rots result in death of the host by desiccation and stem rots can cause water shortage in the shoot despite an abundance in the root. *Pythium* is a classic example of this syndrome: infection occurs at around ground level and collapse of the host tissue is so complete that the seedling is often unable to support itself. Once again, this is a disease particularly prevalent in wet soils and under humid conditions, hence its common name 'damping off'.

Wilting is therefore a symptom of a wide range of diseases ranging from those doing gross damage to the stem and root, through those that interfere with water transport to those that cause wilting by increased water loss.

6.3   Reduction in photosynthesis

Perhaps the most damaging aspect of a large number of infections is the reduction in photosynthesis of infected plants. This is not always apparent and if there is no sign of chlorosis it may simply appear that the plant lacks vigour and is not thriving.

In many diseases the reduced photosynthesis of the host has an obvious cause; in rusts, for example, the death of the leaf tissue caused by the production of fungal fruiting bodies results in reduced photosynthetic area. In wilt diseases the closure of stomata and the resulting limitation of gas exchange will reduce photosynthesis. The breakdown of chlorophyll and the chlorosis that then occurs in some diseases is an equally obvious cause of reduced photosynthesis but the reason for this breakdown of chlorophyll is not so obvious.

Much of the early work on the photosynthesis of diseased plants has been concerned with the breakdown of chlorophyll by the enzyme chlorophyllase, which catalyses the breakdown of chlorophyll to the prophyrin ring containing the magnesium atom (chlorophyllide) and the long side chain (phytol). Accompanying the loss of chlorophyll is the breakdown of the chloroplast structure, the lamellar structure being completely destroyed. The present view is that the breakdown of the lamella is the result of chlorophyll destruction, not the cause of it. The presence of intact chlorophyll molecules is thought to be essential for the maintenance of the lamellar structure.

A number of external factors can influence the degeneration of chloroplasts in diseased plants, one of the most important being the availability of nitrogen. When tobacco is infected with tobacco mozaic virus there is a considerable loss of photosynthetic ability within about nine days. However, if large amounts of nitrogenous fertilizer are applied the symptoms, including loss of photosynthetic

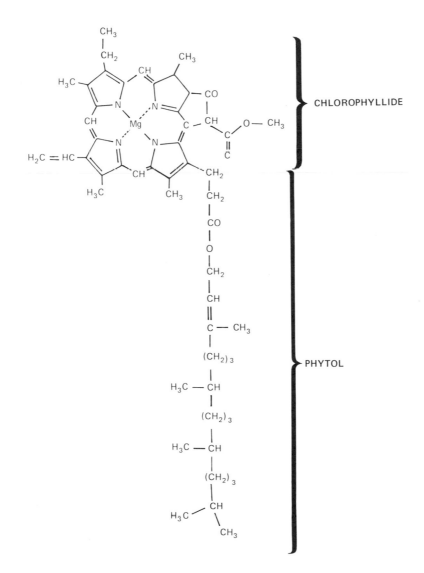

Figure 6.4 Chlorophyll α.

ability, are suppressed. In one experiment the application of nitrogenous fertilizer doubled the rate of photosynthesis of healthy plants but more than quadrupled the photosynthesis of TMV-infected plants, making the rate of photosynthesis of an infected plant with an abundant supply of nitrogen greater than that of a healthy plant having a limited nitrogen supply. The application of nitrogenous fertilizer also reduced the rate of chlorophyll breakdown caused by infection and increased the efficiency of photosynthesis of the chlorophyll molecules to greater than that of an uninfected plant not supplied with fertilizer.

The fertilizer also stimulates a four-fold increase in viral synthesis. Perhaps this is not surprising when one considers that viral synthesis is entirely dependent on the host metabolism and an increase in the metabolic rate of the host is also likely to increase viral synthesis.

Table 6.1 *The effect of nitrogen fertilizer on photophosphyorylation by TMV-infected tobacco leaves.* (Zaitlin and Jagendorf, 1960)

| Treatment | Hill Reaction rate | | | |
	mg chlorophyll /g fresh wt	μmoles $Fe(CN)_6^{3-}$ reduced/g fresh wt/hr	μmoles $Fe(CN)_6^{3-}$ reduced/mg chlor./hr	mg virus/g fresh wt
Uninfected+N	0·79	220	279	—
Uninfected−N	0·57	112	197	—
Infected+N	0·68	182	269	6·3
Infected−N	0·35	42	120	1·4

An interesting phenomenon may be observed in rust infections of wheat and beans. The whole leaf becomes chlorotic with the exception of a small area round the infection site, producing what has been described as a green island. This area is photosynthetically active and starch tends to accumulate there. Work on the powdery mildew of wheat, which displays a similar effect, has shown that the chlorophyll actually reforms in this area following the general destruction involved in chlorosis. The cause of these green islands is not certain but application of kinetin to senescent leaves produces a similar effect, so the cause is probably hormonal.

When bean plants are infected by rust the infected leaves have a reduced photosynthetic activity but the uninfected leaves show a rate of photosynthesis nearly 60 per cent greater than comparable leaves on healthy plants. This will of course partly compensate for the loss of photosynthetic activity of the infected leaves. It is likely, however, that this phenomenon occurs only when the plants are growing under near optimal conditions in which the photosynthetic ability of the leaves on healthy plants is more than able to keep up with the rest of the plant's demand for carbohydrates. In less than optimal conditions it is unlikely that there would be this untapped photosynthetic ability available to make up for the losses in infected leaves.

Chlorosis induced by viral infections does not seem to be quite the same as that caused by fungal and bacterial infections. Most viral infections causing chlorosis not only cause the destruction of some of the chlorophyll but also reduce the efficiency of the residual chlorophyll. Fungal and bacterial infections do not generally affect the efficiency of photophosphorylation, simply causing reduced photosynthesis by destruction of the host's chlorophyll.

Starch metabolism is also affected by the destruction of the chloroplast. Starch may be synthesized in a number of ways. It can be synthesized by the enzyme phosphorylase which polymerizes glucose-1-phosphate (G-1-P) to amylose, liberating inorganic phosphate, but to produce this reaction exceedingly high concentration of G-1-P are required and it is more common for the reaction to proceed in the reverse direction producing G-1-P from amylose. The majority of starch synthesis is thought to be from the sugar nucleotides uridine diphosphoglucose and adenosine diphosphoglucose. It is possible that both the phosphorylase and the sugar nucleotide pathways are co-ordinated because formation of starch from sugar nucleotides cannot take place without at least a tiny amount of preformed starch. The phosphorylase pathway requires no starter molecules of starch and may provide the primer for massive starch synthesis via sugar nucleotides. There is a further enzyme involved in starch synthesis, called the Q enzyme, which forms the 1-6 branching linkages in the formation of starch, the straight chains of which (amylose) are 1-4 linked glucose molecules.

It is not clear where in these pathways the pathogens cause a disturbance. TMV-infected tobacco leaves fail to synthesize and to degrade starch at normal rates. Infected plants taken at the end of a period of active photosynthesis contained less starch than healthy plants but after a prolonged period of darkness less starch had been hydrolysed in the leaves of infected plants than in healthy plants; as a result of this some infected leaves contained more starch after these treatments than similarly treated healthy leaves.

Starch accumulation is a characteristic of a number of other virus diseases, such as potato leaf roll, sugar beet yellows and barley yellow dwarf. In some instances the accumulation of starch may be due to phloem necrosis resulting in the plant's inability to translocate the sugars away from the leaf, but in sugar beet yellows phloem necrosis does not occur. Infection of sugar beet by yellows virus does not in any way impair the plant's ability to translocate sugars and yet there is a build-up of carbohydrates in the leaf. Neither does phloem necrosis explain the situation where there is starch depletion in yellowed areas of a chlorotic leaf and yet only a few cells away intact chloroplasts are accumulating starch.

Similar metabolic abnormalities are found in bacterial and fungal infections which cause damage to chloroplasts but no general hypothesis has been put forward to explain the observed effects in these diseases. The enormous demand for nucleic acids during viral synthesis has been suggested as a possible cause of reduced starch synthesis in viral infections but this does not explain the reduced starch breakdown and the greatly slowed movement of carbohydrates, nor does it explain reduced starch synthesis in fungal and bacterial infections. Some workers have found reduced phosphorylase activity in potato leaves infected with potato virus X. This could be the cause of reduced breakdown and possibly of reduced synthesis but it would not explain local accumulation.

The disturbance of starch metabolism seems to be associated with damage to the chloroplast but it is likely that the different ways in which it is disturbed are due to different processes, and it is probable that this varies not only with the type of disturbance but also with the type of disease. It is to be hoped that as more experimental data become available a pattern will emerge that will enable us to group together diseases which damage the host in the same way and will therefore respond to the same treatment.

6.4   Alterations in respiration

The purpose of respiration is two-fold: to provide energy for the various metabolic processes in the cell, and to provide some of the simple organic molecules required in the various synthetic pathways. Respiration is usually considered to start with glucose which, under aerobic conditions, is oxidized to carbon dioxide and water, and under anaerobic conditions to carbon dioxide and ethanol or organic acids. Other storage materials, such as starch or fats, could just as logically be considered as the starting point of respiration. The ratio of carbon dioxide liberated to oxygen utilized varies depending on the food reserve being oxidized. For carbohydrates the equation is:

$$C_6 H_{12} O_6 + 6O_2 \rightarrow 6 CO_2 + 6H_2 O$$

Thus the respiratory quotient (RQ) which is the ratio of carbon dioxide produced to oxygen used, is equal to 1. When fats are oxidized, as in the germination of seeds or spores, the respiratory quotient is about $0 \cdot 7$, and for compounds richer in oxygen, such as malic acid, it can be as high as $1 \cdot 3$.

$$(C_4 H_6 O_5 + 3O_2 \rightarrow 4CO_2 + 3H_2 O)$$

This difference in respiratory quotient provides the physiologist with some idea of the food reserve being used.

During the breakdown of these respiratory substrates considerable amounts of ATP are produced from adenosine diphosphate (ADP) and inorganic phosphate (P_i). ATP is the energy-carrying intermediary used in the majority of synthetic processes in the cell. The energy theoretically available from the aerobic breakdown of glucose is around 640 Kcal/mole but from anaerobic breakdown it is only 54 Kcal/mole. This is reflected in the amounts of ATP synthesized per molecule of glucose degraded. Under aerobic conditions about 36 ATP molecules are produced per molecule of glucose oxidized, 8 from glycolysis and 28 from the Krebs cycle. Under anaerobic conditions however the energy from Krebs cycle is not available and as the conversion of pyruvic acid to ethanol uses 6 of these 8 ATP molecules, the net gain per molecule of glucose is only 2 ATP. This shows that not only is less energy available under anaerobic conditions, it is utilized less efficiently in the synthesis of ATP. The theoretical ratio of aerobic to anaerobic energy available is approximately 12 to 1 but the actual ratio of ATP produced is 18 to 1.

As much more energy is produced under aerobic conditions the rate of breakdown of sugars need not be as fast in the presence of oxygen as in its absence. The rate of respiration seems to be largely controlled by the rate of glycolysis and this in turn is controlled by the ratio of ATP to ADP. The control of glycolysis, reducing its rate in the presence of oxygen, is called the *Pasteur effect* after its discoverer. Metabolic disorders can however upset this control. Such a disturbance can be produced by treating plant tissue with DNP which uncouples respiration from ATP synthesis (i.e. oxidative phosphorylation) leaving the cell short of ATP. The cell responds by increasing the rate of respiration to its maximum in an endeavour to maintain ATP levels. Oxygen will now have no effect on the rate of respiration which is at an absolute maximum.

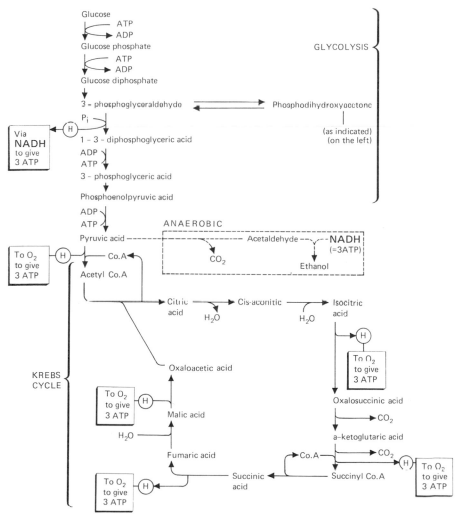

Figure 6.5 The respiratory sequence of glycolysis and Krebs cycle showing the energy production.

A wide range of plants show increased respiration in response to almost any form of trauma, including mechanical and chemical injury as well as infection. The increase in respiration caused by fungal infections usually coincides with the appearance of visible symptoms and rises to a maximum at about the time of sporulation. It is not clear how much the pathogen contributes towards this increased respiration but RQ measurements suggest it is largely due to the host. The RQ is usually very close to 1·0 even in infections caused by fungi in which oil is the major food reserve.

The increased respiration might conceivably be due to the uncoupling of oxidative phosphorylation (the same sort of effect as DNP) or it could be caused by accelerated host metabolism. A third possibility is the breakdown by the pathogen of ATP to ADP and P_i, during which it may or may not be able to utilize the liberated energy. The evidence suggests that most of the increased respiration is due to increased host metabolism and there have been many reports of increased growth, protein synthesis and nucleic acid synthesis as a result of infection. Not only is the rate of respiration raised but the temperature of the diseased tissue rises as high as 0·7 deg C above that of a healthy control plant.

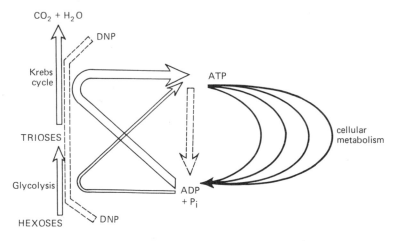

Figure 6.6 The production and utilization of ATP in a cell. Indicated is the way respiration can be uncoupled from ATP synthesis and how ATP breakdown can occur outside the metabolic control of the host cell.

There is little evidence about the efficiency of diseased respiration but there are some indications that it is lower than in healthy tissue. This may be responsible for the rise in temperature of infected tissue. An example of this apparently lower efficiency is that the oxygen consumption per unit of nucleic acid synthesized is greater in rust infected wheat than in healthy wheat. The drop in efficiency is possibly related to the disappearance of the Pasteur effect (oxygen inhibition of glycolysis) in some tissues following infection, as for example wheat infected with rusts and sweet potato infected with *Helicobasidium mompa*. We have already noted that anaerobic respiration is less efficient than aerobic respiration and it has been suggested that the drop in efficiency of the respiratory system is due to aerobic and anaerobic respiration occurring simultaneously in infected tissue. Clearly this is not the entire explanation because even a small amount of anaerobic respiration would dramatically alter the respiratory quotient. There is no evidence of any such change in respiratory quotient nor is there a build-up of ethanol or organic acids following infection. Perhaps anaerobic respiration does take place and the waste products of this process are further oxidized to carbon dioxide and water so that there is no obvious change in respiratory quotient or build-up of waste metabolites, but the overall efficiency of this process would nevertheless be lower than normal aerobic respiration. The oxidation of anaerobic waste products can certainly occur in some instances, as for example when oxygen is made available to

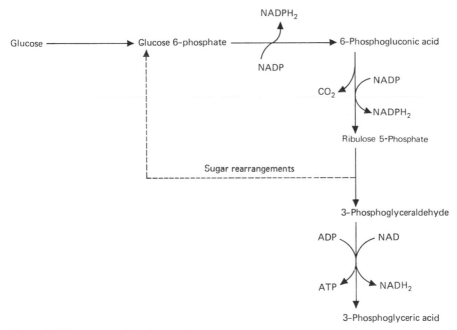

Figure 6.7 The pentose phosphate pathway.

healthy plant tissue which had previously been respiring anaerobically in the absence of oxygen.

There is a route other than glycolysis by which glucose can be broken down to 3-phosphoglyceric acid: the pentose phosphate pathway. There is good evidence that the pathway is of considerable importance in the metabolism of many diseased plants. If glucose is broken down via the pentose phosphate pathway most of the carbon dioxide will be evolved first from the C-1 position, whereas if respiration proceeds via glycolysis carbon dioxide will be released equally rapidly from the C-1 and C-6 positions. Use of ^{14}C glucose has enabled the importance of each pathway to be determined in healthy and diseased plants, and the results suggest that an increase in the pentose phosphate pathway activity accounts for the increased respiration of diseased plants. Certainly in wheat, bean and sunflower infected with rust and in barley infected with mildew there is an increase in the proportion of C-1 carbon in the carbon dioxide evolved at the time of sporulation of the pathogen, time of maximum respiratory stimulation.

Increased respiration rate may be part of the host defence mechanism. Generally, plants resistant to fungal infection show a greater increase in respiration than susceptible plants, and narcotics which suppress respiration tend to increase the susceptibility of plants to fungal infection. Ethanol also suppresses respiration as well as the synthesis of proteins and phenolics in potatoes and decreases their resistance to infection. Similarly the loss of ATP by uncoupling agents such as DNP also decreases resistance to infection.

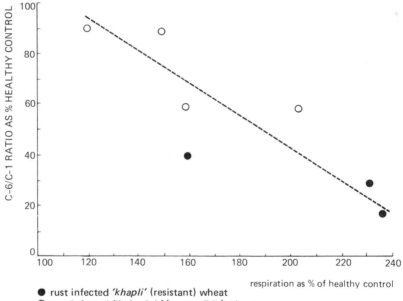

● rust infected *'khapli'* (resistant) wheat
○ rust infected *'little club'* (susceptible) wheat

Figure 6.8 Showing the change in the C-6/C-1 ratio of $^{14}CO_2$ from plants fed $[^{14}C\text{-}1]$ glucose and $[^{14}C\text{-}6]$ glucose as respiration is stimulated by infection. Also shown is the greater stimulation of respiration in resistant than in susceptible plants.
(after Shaw and Samborski, 1957)

The relationship between resistance and respiratory stimulation by bacterial infections is even less clear cut. Early in infection susceptible plants frequently show far greater respiratory stimulation than resistant plants but the respiration of the susceptible plants then tends to decline whilst that of resistant plants steadily increases. Often there is an increase in respiration following viral infection but this is not universal and a few viral infections even cause a decrease in the respiration of the host. Any stimulation of respiration again tends to be late in the disease cycle, in local infections it reaches a maximum at around the time of the first visible symptoms and in systemic infections it reaches a maximum even later when symptoms have been visible for some time.

As respiration of a plant is stimulated by mechanical damage, and inoculation with virus frequently involves some degree of mechanical damage, even greater care than usual must be taken to ensure the control plants truly differ only by the absence of the virus. This means that the same damage must be inflicted on the control plants as that produced by the inoculation process. One problem in measuring the rate of respiration of a plant is that different tissues in the same organ respire at different rates and even in comparatively homogeneous tissues local infections will alter the rate of respiration but may leave the respiration of the surrounding tissue unaffected, thus causing a very tiny overall change in respiration.

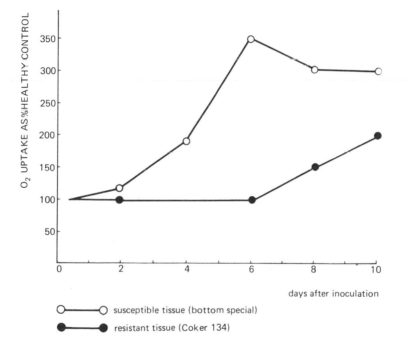

days after inoculation

○────○ susceptible tissue (bottom special)
●────● resistant tissue (Coker 134)

Figure 6.9 Stimulation in respiration of tobacco by a bacterial infection (*Pseudomonas solanacearum*). Initially the susceptible plants show a greater stimulation of O_2 uptake than do the resistant plants.
(after Maine, 1960)

As viral synthesis is entirely dependent on the host metabolism it is to be expected that any treatment upsetting the latter will cause a decrease in viral synthesis. Respiratory inhibitors which tend to increase susceptibility to fungal infection will therefore slow down viral synthesis, and conversely, application of respiratory intermediaries, such as tricarboxylic acids, will tend to cause an increase in viral synthesis. Experimental findings show that these suppositions are subse-

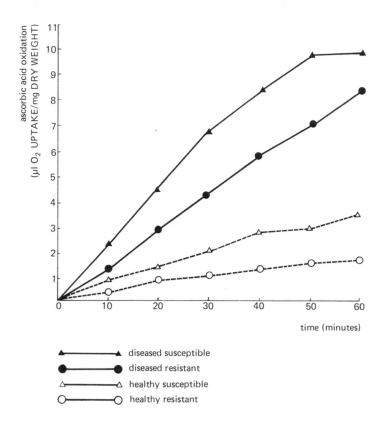

Figure 6.10 Ascorbic acid oxidation by the enzymes in homogenates from resistant (Coker 139) and susceptible (Bottom Special) tobacco plants 13 days after inoculation with *Pseudomonas solanacearum.*
(after Maine, 1960)

quently correct but although the rate of viral synthesis is altered by these treatments they have comparatively little effect on the initial resistance or susceptibility of the plant to infection.

Although most of the increased oxygen uptake by an infected plant is associated with respiration it must be remembered that there are other oxidative processes stimulated by infection which also form part of the host's defence mechanism, and these include the oxidation of polyphenols to melanoid pigments.

6.5   Disturbance of growth regulators

One of the substances most extensively and intensively studied by plant physiologists is auxin (indolylacetic acid — IAA). It has been investigated both in healthy and diseased plants, and marked increases in concentration have frequently been found following both bacterial and fungal infections. In fact, raised respiration and increased auxin content are among the commonest of the physiological symptoms of infection. The increased auxin content of the tissues might be due to: (i) increased production by the host; (ii) production by the pathogen of auxin or auxin precursors (such as tryptophan); (iii) reduced auxin breakdown by the host; (iv) inhibitors of auxin breakdown produced by the pathogen.

Tobacco plants infected with *Pseudomonas solanacearum* have been found to contain over ten times the concentration of auxin found in healthy plants (Healthy; 0·3 μg IAA/100 g tissue. Diseased; 3·3 μg IAA/100 g tissue). The hormone was present in a concentration too low to be detected regularly in the stems and roots of healthy plants with the maximum concentration occurring in the leaves, whereas in infected plants the maximum auxin concentration was in the stem and roots where the pathogen grows most actively. This does not prove the hormone was synthesized by the bacterium. It may be produced by the host in response to the pathogen or the bacterium may cause a reduction in the rate of auxin breakdown. More direct evidence for the synthesis of auxin by the pathogen is available in the case of *Agrobacterium tumefaciens* infections where intermediaries in the pathway converting tryptophan to auxin have been isolated from pure cultures of the organism. Similarly *Pseudomonas savastanoi* has been found to be capable of converting tryptophan to auxin in pure culture. Many fungal infections also result in increased auxin concentrations in the infected plant. Following the infection of hypocotyls of safflower seedlings with *Puccinia carthami* the auxin level reached a maximum about twelve days after inoculation at which time it was over four times the concentration found in healthy hypocotyls; after this the concentration started to decline. These elevated auxin levels resulted, as would be expected, in an increased enlargement of the hypocotyl cells causing an elongation of this part of the seedling.

There are considerable variations in the auxin content of different tissues of a host and in hosts of differing resistance to infection. Susceptible potato tissue infected with *Phytophthora infestans* had five to ten times the auxin content of healthy tissue but in resistant potatoes the auxin was inactivated in the infected region resulting in subnormal auxin concentrations. Just below the infected tissue, however, there was a fifteen- to twenty-fold increase in the concentration of auxin compared with healthy control tubers.

That an organism is capable of synthesizing auxin in pure culture does not necessarily mean that this will occur when it is growing in the host. Synthesis of auxin may be limited by the availability of a suitable substrate. There are indications that higher plants and micro-organisms convert tryptophan to auxin by different pathways so it should be possible to determine from the intermediaries present whether production is by the host or the pathogen.

Hyperauxinity can be produced not only by increased synthesis but by inhibited breakdown. As auxin is thought to be degraded largely by IAA oxidase, inhibition of this enzyme would result in a build up of auxin. IAA oxidase is severely inhibited in the rust disease of *Euphorbia* caused by *Uromyces pisi*. In the early stages of the disease there is a very close correlation between decrease in IAA oxidase activity and the increase in auxin content of the tissue but later in the disease the reduced destruction of the hormone and its continued production cause fairly high concentrations of auxin to be reached.

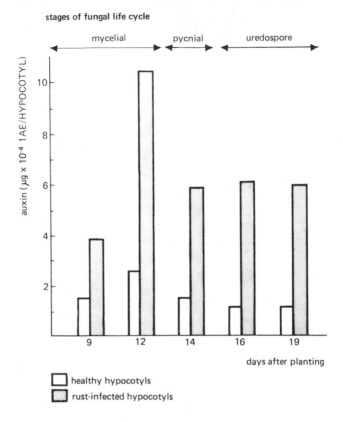

Figure 6.11 Increased IAA content of Safflower hypocotyl as a result of rust infection (after Daly and Inman, 1958)

Table 6.2 *Auxin content and auxin destruction in healthy stems of* Euphorbia cyparissias *and in stems infected with* Uromyces pisi. (after Pilet, 1960-1)

State of plant (healthy of same age as diseased at stage indicated	IAA content in µg IAA equivalent/kg fresh wt, at following stages:		IAA destruction in µg IAA equivalent destroyed/min/kg fresh, wt at following stages:	
	Mycelium	Aecidiospores	Mycelium	Aecidiospores
Diseased	22·8	39·7	1·68	1·76
Healthy	12·4	7·8	3·17	3·47

Work on the nature of the IAA oxidase inhibitor has shown that the inhibitory substance produced in tobacco roots infected with *Pseudomonas solanacearum* is of small enough molecular weight to pass through a dialysis membrane. Dialysis of extracts from infected roots increased the ability of the extract to oxidize IAA by a factor of nearly six but dialysis of extracts from healthy plants only increased their IAA oxidase activity by about 0·5. A number of water-soluble phenols were detected in the dialyzate, scopoletin being most concentrated. It is quite possible that scopoletin is the IAA oxidase inhibitor. Experiments *in vitro* show it is capable of suppressing the enzymic degradation of auxin and it is present in diseased tissue at up to ten times the concentration in which it is found in comparable healthy tissue. Whether scopoletin is the inhibitor or not it is still not clear if the IAA oxidase inhibitor is produced by the pathogen or by the host in response to the pathogen.

The reverse of these effects is found in some diseases in which there is stimulation of IAA degradation, as a result of which the auxin concentration falls. This fall may well be the cause of premature leaf drop by some infected plants. Certainly disease-induced leaf fall can often be delayed by the application of auxin or auxin analogues to the infected plant.

As well as acting as a palliative in some diseases exogenously applied auxin actually increases the resistance of a few plants to infection: the application of high levels of auxin to tomato plants increases their resistance to *Fusarium* wilt, and of potatoes to *Phytophthora* blight. These effects may be due to the involvement of hormones in cell wall growth. Treatment of tomato plants with the auxin analogue naphthalene acetic acid increased the calcium bonding in the cell wall pectin, thereby reducing its solubility. The hydrolysis of pectin may be involved in the vascular wilt syndrome as a consequence of the pectin hydrolysate increasing the viscosity of the xylem sap thereby reducing the flow rate of the transpiration stream and resulting in a water shortage in the leaf. The degree of calcium bonding could therefore be important in determining the rate of hydrolysis of the pectin and thus the effect of the pathogen on water transport to the leaf.

The application of exogenous auxin is not always beneficial in the treatment of plant diseases. High auxin levels seem to increase the susceptibility of tobacco to *Pseudomonas solanacearum,* the cause of Granville wilt. There are a number of suggested possible causes of this increased susceptibility, one being that the auxin causes greater cell wall plasticity as a result of increased metal binding and reduced cross linking between the pectin chains. This causes the tobacco tissues to become more succulent with a higher water content and therefore more easily attacked by the bacterium. A second suggestion is that the increased susceptibility is due to reduced lignification while a third is that the plant is more affected by the pathogen because the auxin causes increased transpiration. None of the three explanations are mutually exclusive and all three effects could be operating simultaneously.

Whatever the source of the auxin, and whether or not infection results in an increase or decrease in the auxin content of the host tissue, these hormonal disturbances are frequently the most obvious of the disease symptoms. They can result in a whole variety of growth abnormalities such as outgrowths of unspecialized callus tissue, twisting and distortion of stems and leaves, dwarfing of leaves and stems, enlargement of stem internode lengths, production of adventitious branches and roots, etc.

Table 6.3 *Pathogenic relationships and their physical manifestations in the plant.*

Symptoms indicative of hormonal disturbance	Disease	Host	Pathogen
Lengthening of internodes	rust	wheat	*Puccinia graminis*
	rust	safflower	*Puccinia carthami*
	rust	*Capsella*	*Albugo candida*
	rust	*Euphorbia*	*Uromyces pisi*
	smuts	barley	*Ustilago hordei*
	smuts	wheat	*Ustilago tritici*
	smuts	*Bromus*	*Ustilago hypodytes*
	downy mildew	sugar cane	*Sclerospora sacchari*
	bakanae disease	rice	*Gibberella fujikuroi*
Stunting	yellow dwarf	barley	Barley yellow dwarf virus
Apical stunting	cauliflower disease	strawberry	*Corynebacterium fasciens*
Stem distortion	downy mildew	*Brassicae*	*Peronospora parasitica*
	rust	thistle	*Puccinia suaveolens*
	rust	groundsel	*Puccinia lagenophorae*
Leaf distortion and premature leaf fall	leaf curl	peach	*Taphrina deformans*
Leaf callus and adventitious shoots	leaf gall	sweet pea	*Corynebacterium fasciens*
	leaf gall	chrysanthemum	*Corynebacterium fasciens*
Leaf necrosis but abscission prevented	leaf scorch	cherry	*Gnomonia erythrostroma*
Areas of leaf necrosis and premature leaf fall	leaf spot	coffee	*Omphalia flavida*
	black spot	roses	*Diplocarpon rosae*
Collar or root callus	crown gall	many	*Agrobacterium tumefaciens*
Root callus	club root	*Brassicae*	*Plasmodiophora brassicae*
Tubercle on young leaves and twigs	olive knot	olive	*Pseudomonas savastanoi*
Callus on twigs	cedar apple rust	red cedar	*Gymnosporangium juniperi-virginianae*
Callus on tuber	wart	potato	*Synchytrium endobioticum*
Epinasty and adventitious roots	vascular wilt	tomato and legumes	*Verticillium albo-atrum* *Fusarium oxysporum*

Auxins are not the only plant hormones implicated in the host-parasite inter-action. Cytokinins appear to be involved in the diseases causing chlorosis. The chlorotic symptoms can be suppressed in many instances by the application of kinetin and the protein degradation associated with these diseases can also in many instances be reversed by the foliar application of kinetin. In *Pseudomonas tabaci* infections the effects of cytokinins seem to be due to the inactivation of a toxin produced by the pathogen which inhibits methionine incorporation into protein. Toxin inactivation is clearly not the only effect of exogenous cytokinins as they can delay chlorosis in senescent but uninfected leaves and delay the abscission of such leaves.

The production of green islands of photosynthetic and metabolically active cells around sites of rust infection, whilst the rest of the leaf tissue shows symptoms of chlorosis, bears all the hallmarks of locally increased cytokinin production by either the host or pathogen. Other examples of cytokinin-like action following infection include the stimulation of large numbers of lateral buds in sweet peas infected by *Corynebacterium fasciens* and then the inhibition of their elongation, and also tumour development such as occurs in oleander infected with *Pseudomonas savastanoi.*

In viral infections the role of cytokinins is less apparent and there is considerable variation between the findings of different workers, perhaps dependant on the physiological state of the host. Low concentrations of kinetin stimulate tobacco mozaic virus synthesis in *Nicotiana glutinosa* leaves but high concentrations of kinetin have been shown to inhibit TMV synthesis both in entire leaves and leaf discs of *N. glutinosa.* However a wide range of concentrations of kinetin have been found to stimulate the synthesis of tobacco necrosis virus in entire french bean plants. Conversely the infection of tomato and petunia by tomato spotted wilt virus is reduced by the application of kinetin.

Experiments with detached leaves and leaf discs have the advantage that the source of unknown amounts of endogenous hormones has been removed but it has the disadvantage that the tissues will now have grossly abnormal hormone, sugar and water balances. The application of kinetin may increase viral synthesis simply by delaying senescence of the host tissue and in other instances it may reduce viral synthesis by diverting metabolites away from viral synthesis towards maintenance of the host cell. This might account for the variation in experimental results.

Gibberellins too are involved in the host-parasite interaction. The effects of gibberellic acid were first observed in the bakanae disease of rice caused by *Gibberella fujikuroi.* The fungus infects the roots of the rice seedling and produces gibberellic acid which greatly stimulates the growth of the shoot, but the plants soon die due to the complete destruction of the roots by the fungus. This rapid growth of the seedling followed by collapse and death leads to the Japanese name for the disease, 'foolish seedling'. The active component produced by the fungus was isolated and named after the genus of fungus producing it. Since then many slightly different gibberellic acids have been isolated from a number of different micro-organisms and higher plants.

Ethylene is another very active growth regulator with effects very similar to auxin. The detection of this compound on a number of occasions at the site of infection or wounding suggests that it may be more important in the host-parasite relationship than has previously been thought. Normal fruit ripening is often associated with ethylene production so it is perhaps not surprising that increased ethylene concentrations have been found in fruit showing premature ripening as the result of infection. The increased ethylene content of vegetative tissue that is some-times present as a result of infection is rather more unexpected. Epinasty (the downward bending of leaf petioles) is a common symptom of both high auxin and high ethylene content of host tissues and is seen in a number of wilt diseases. Work has also been carried out on plants showing local lesions as the result of virus infection. They too were found to have a raised ethylene content.

The two main hypotheses put forward to explain these high levels of ethylene are: (i) ethylene production by the pathogen; and (ii) closure of stomata reducing the release of host-produced ethylene during transpiration. Ethylene production has been recorded for a number of pathogens growing in artificial culture and the production of the gas by bacterial pathogens seems to be a fairly common phenomenon. Ethylene is usually lost very rapidly from healthy tissue during transpiration so that reduced transpiration from whatever cause would favour the build-up of ethylene whether it was produced by the host or pathogen. Increased ethylene content of plant tissue is often associated with increased auxin content and as their effects are similar it is difficult to determine which is the cause of the observed symptoms and quite likely that both are involved. For instance, tomato infected with *Verticillium albo-atrum* (sleepy disease) shows wilt symptoms, but epinasty and sometimes leaf curl take place, as well as the production of adventi-tious roots just above soil level. These symptoms might follow from ethylene build-up resulting from closed stomata, or from the raised auxin concentration that has been detected.

As more work is carried out on the physiology of healthy plants it becomes more and more apparent that the interaction of plant hormones and the balance between them is often of far greater importance than their absolute concentration. Gibberellic acid, for example can stimulate internode elongation only in the presence of auxin; i.e. the two hormones act synergistically. It is therefore likely that the symptoms one sees in diseased plants are the result of a hormone imbalance, often compounded by some hormones increasing in concentration as a result of infection whilst others decrease. This is difficult to demonstrate conclusively because of the multiplicity of effects that a single hormone can have, and acting in conjunction their effects are probably even more wide ranging than we realize at present. The steady build-up of experimental data is nevertheless helping us to move towards a more unified understanding of the physiological interaction of host and parasite.

Chapter 7
Disease control

The aim of any agriculturalist is to grow disease-free plants and it is to this goal that the whole study of plant pathology is aimed. There is, however, no point in growing disease free plants in this way if, as a result of maintaining them free of disease, the final yield is lower than it would be under normal commercial conditions. Resistant plants would hardly be grown if their maximum potential yield was less than the yield of high-yielding varieties suffering from average levels of infection, and cultural practices would not be employed to reduce infection if it resulted in a lower yield. The aim of reducing disease is to increase yield and this must always be borne in mind. The reduction of infection is not an end in itself but merely a means to an end and that end must always justify the means of disease control employed.

7.1 Control by cultural practice

It is fairly obvious that if one is to grow uninfected plants, disease-free seeds must be sown. This is not always easy to do, and it is usually impossible to recognize infestation from superficial observation. The needs must therefore be tested for freedom from disease by one, or, better still, all of the following methods:

(i)   superficial observation, both microscopic and macroscopic, of seed samples and the microscopic observation of seed parts such as the embryo;
(ii)   Attempts to culture possible pathogens from the seed on a variety of media;
(iii)   Germination of the seed and growing the resulting plants under conditions in which they will not contract the disease from outside.

There are problems with all the procedures. Superficial observation will detect only those diseases with large amounts of mycelium or spores in the seed, and the few diseases causing gross abnormalities in the seed's development. The majority of seed-borne diseases will not be detected by this method. Culture of the possible pathogens carried by a seed will not detect highly obligate parasites which can only grow in a living cell. Growth of the seed may not even show the presence of a pathogen as it may only develop under particular environmental conditions or only very slowly. Or again it might become apparent only when the host reaches maturity, which makes it impossible to test such slow maturing species as trees in this way. A problem associated with all screening methods is that of sampling the seed for testing. Any infected seed is unlikely to be uniformly distributed as it probably came from a particular part of a field being harvested and would therefore be restricted to particular sacks or parts of a sack. This clearly poses problems of sampling which can only be partially overcome by statistical analysis of a large number of samples.

All these remarks about seed testing also apply to other material planted to start a crop, such as bulbs, corms, cuttings, 'seed' potatoes and other tubers, all of which are more prone to carry disease than true seeds and must therefore be screened even more carefully.

It is obviously not practicable for a farmer to carry out these procedures himself and the seed producers therefore subject the seed to a critical examination to enable it to be certified as reasonably free from disease. Even seed producers cannot guarantee absolute freedom from all diseases because of the testing problems already mentioned, but to reduce further the likelihood of the seed carrying serious diseases it is common for the seed producers either to grow the seed themselves in areas where serious diseases of that crop are not endemic (or to have the seed grown for them by farmers in such areas), and to inspect the crop and the crops of surrounding farmers to make sure the parent crop is healthy and is not in contact with infected ones.

Sowing conditions can alter the likelihood of a seed or seedling being attacked by a pathogen. Mechanical damage to seeds or to vegetative organs during sowing will predispose them to attack by micro-organisms present in the soil. (Seeds with very tough testas are occasionally deliberately damaged or scarified to increase the rate of water penetration.) Depth of sowing can also be important. Generally it is advisable to sow seed at as shallow a depth as will avoid desiccation. This is because the vigour of the germinating seed is greatest under aerobic conditions and the greater the vigour of the seed the greater its resistance to infection. Very occasionally deep sowing can be useful, for example when perennial rye grass seed is known to contain a percentage of seed infected with *Gloeotinia temulenta* (the cause of blind seed disease). The infected seed does not germinate but the fungus carried in it produces an apothecium which it pushes up above the soil and releases spores, infecting the surrounding plants. If the seed is sown more than about 2-3 cm below the surface the apothecia do not reach the surface and the surrounding plants are not infected.

The sowing date is an important consideration. Usually a date is chosen that is optimum for the vigour of the crop but sometimes dates before or after this are chosen, perhaps to produce an exceptionally early or late crop, when a good price will be obtained. A sowing date other than the apparent optimum may then be used to avoid the crop being at a susceptible stage in its development when a particular disease is prevalent. Cases in point are the early sowing of potatoes to enable growth to be practically complete before *Phytophthora* blight is a problem, and the early sowing of broad beans to reduce the problems of aphid infestation. Alteration in sowing times can however lead to other problems. As we have already mentioned, anything which reduces the vigour of a plant will tend to increase the likelihood of it succumbing to an infection, so that sowing at other than optimum times can increase the chances of disease. Climatic conditions can also damage crops grown at unusual times of the year. Early or late frosts can damage both the vegetative parts and the blossom of plants, and so can drought or excessive rain at a critical stage in the plant's development.

It is fairly generally accepted that high levels of fertilizer, particularly nitrogenous fertilizers, as well as increasing crop yield can also increase the susceptibility of the crop to some infections. The effects of mineral nutrition have already been discussed but in brief the nitrogenous fertilizers increase susceptibility to such diseases as rusts, powdery mildews and blasts but have little effect on quite a large number of other diseases; it has even been reported that it can reduce the likelihood of infection by smut fungi. Potassium and phosphate fertilizers tend to increase crop resistance to infection but here again there are exceptions, and virus infections can in some instances be made more serious by the application of these compounds. The type or types of fertilizers chosen are based on the mineral requirements to bring the soil into optimum condition for growing a particular crop. It is however worth while also to consider the effect the fertilizer is likely to have on disease, particularly those diseases prevalent in the area. In this way a higher yield may be obtained by reducing the loss due to disease even though the mineral constituents of the soil may be suboptimal for the crop.

Trap crops are used to bring about the germination of pathogenic propagules but then to cut short their lifecycle so that no new propagules are produced, and over a period of time soil can be cleared of a pathogenic infestation. This is usually brought about either by the sowing of a crop that stimulates germination of the pathogen but which is resistant to its attack and therefore does not become a host or alternatively, if such a crop is not available, a susceptible crop is sown and then ploughed in as a fertilizer or cut for silage before the life cycle of the pathogen is complete. For example, witchweed (*Striga*) a flowering plant parasite of cereals in warm climates can sometimes be controlled by sowing a grass such as Sudan grass and then ploughing it in before the witchweed sets seed.

Trap crops are only likely to be effective against highly specialized obligate parasites that will germinate only in the presence of a suitable host. They are unlikely to be successful against less specialized pathogens and attempts to use them may even prove harmful if the pathogen is able to grow saprophytically on the ploughed-in debris.

As pathogenic propagules age they lose their vigour and ability to infect a host and may even die if a suitable host is not available for a considerable period of time. Crop rotation can therefore be used as a control measure. It is carried out partly to avoid the selective depletion of minerals in the soil that would occur with the continuous growth of one particular crop, but it is also effective in reducing disease levels in the soil. It is rarely successful in completely freeing soil of a pathogen because the economic requirements are usually such that a limited number of crops are grown in rotation and this does not allow enough time between similar crops to ensure the death of all the pathogenic propagules. The times required for the death of all the propagules is usually very long because their death tends to cause a logarithmic decline in population so a low level of infestation will be present for a long time. Crop rotation is also of less use against facultative parasites than against obligate parasites because the former can grow saprophytically for considerable periods of time in the absence of the host. Pathogens with a very wide host range will not easily be controlled by crop rotation as it is likely that a number of the crops used will be susceptible. *Phymatotrichum omnivorum,* for instance, attacks over 2000 different species of plants, many of which are common field weeds that would carry the infection even if the crop grown was resistant. The growth of crop pathogens on field weeds is another reason why crop rotation may not be completely successful in controlling infection.

Despite these limitations rotation is good crop husbandry practice in that it reduces the level of infection of many diseases even if it is not able to eliminate them entirely. Crop rotation in common with all other forms of sanitation reduces the original inoculum level at the beginning of the growing season and in the discussion of epidemiology we have seen how this can be an effective means of reducing disease in a crop.

7.2 Chemical control

Treatments for fungal and bacterial infections can be divided into a number of groups (there is as yet no effective chemical treatment for viral infections):

(i)   Protective: those which inactivate the pathogen before it infects the plant;
(ii)   Therapeutic: those which inactivate the pathogen once it has entered the plant. Each of which can be sub-divided into
       (a) fungicidal or bacteriocidal: those which kill the pathogen;
       (b) fungistatic or bacteriostatic: those which stop the growth of the pathogen or inhibit some particular stage in its life cycle, such as sporulation.

Many plant protection compounds act in more than one of these ways, and may act in both simultaneously. They may be fungistatic to some fungal species and fungicidal to others, or may act sequentially as a therapeutic agent and then as a protectant. Alternatively they may act in only one way at a time, depending on the mode of application or the time of application; a compound may for instance act as a protectant if applied before infection, or therapeutically if applied afterwards.

Application of the treatments varies with the crop, the situation, the formulation, the pathogen aimed at and a number of other practical considerations. Protection of the crop may start with soil sterilization. Steam treatment although effective on small samples is hardly practical on a large scale, and chemical sterilization is generally used. In any case, total soil sterilization is rarely required, the destruction of a particular pathogen and therefore a selective or partial sterilization treatment being adequate. Whether total or partial sterilization is the aim the compound used as well as being effective against the pathogen must have certain other properties. To achieve penetration of the soil crumbs the compound should be somewhat volatile but this in itself creates difficulties, making its use expensive and tending to limit it to seed beds, glasshouses and crops of high value. The earliest soil 'fumigants' were formaldehyde and methyl bromide. These are still sometimes used but more modern compounds are taking their place such as *Vapam* (sodium

N-methyldithiocarbamate) and *Dazomet* (tetrahydro-3,5-dimethyl-1,3,5-thiadiazine-2-thione) which break down in the soil to produce a fumigant gas. This makes them easy to apply but their cost makes their use limited. To be effective soil fumigants must be incorporated into the surface soil, usually the top 10-20 cm, either by mechanical mixing (ploughing or harrowing) or by injection under pressure from the tips of special drills. Alternatively they may be washed into the soil with large amounts of water. All these methods clearly add to the cost of the treatment.

Treatment of the seed (seed dressing) is a far more economic method of protecting the germinating seed from attack. A seed dressing should have a high microbial toxicity but a fairly low solubility so that it is not rapidly washed off and yet provides a solution of sufficient concentration to control the growth of possible pathogens. Ideally, too, the compound should have low animal toxicity so that it does not affect wildlife with which it comes into contact.

If the dressing is applied as a powder to the seed it is difficult to get both adequate attachment of the compound and uniform mixing of the seed and dressing. The first difficulty can be overcome by adding to the dressing formulation an inert carrier powder with good surface retention properties, but it is still likely that there will be variation in the amount of dressing different seeds receive. In extreme cases some may get little or no dressing whilst other seeds receive a phytotoxic dose. Despite these difficulties, the dry application of seed dressings is successfully used with a large number of compounds. Seed is also treated by a wet method in which it is soaked or drenched in a solution or suspension of the dressing, and then dried in a blast of hot air. This process gives very uniform dosage but the soaking and drying is cumbersome and expensive. The drying must be carefully controlled to avoid thermal damage to the seed and yet must be complete otherwise premature germination and spoilage would result. Compounds sometimes applied in this way include *Captan, Dichlone, Chloranil, Spergon* and *Thiram*. A compromise treatment between wet and dry application of the dressing is to apply it as a slurry. This gives more uniform coverage than the dry treatment and easier drying than the wet treatment.

Whatever method of application is used good adhesion of the dressing to the seed is essential so that in the subsequent handling the dressing does not come off the seed and accumulate at the bottom of sacks and containers. To this end, water-soluble adhesives are sometimes mixed with the wet- and slurry-applied dressings. Seed dressings have the secondary advantage of reducing seed loss during storage by inhibiting microbial activity. They cannot of course be applied to seed destined for human or animal consumption and a dye is almost invariably incorporated into the dressing to make it obvious that the seed has been treated.

Chemical treatments can be applied to mature plants as well as to seeds. The method of application will depend on several factors including the target pathogen, the chemical selected and the prevailing local conditions. Roots are treated using techniques similar to chemical soil sterilization although with different compounds. The materials can be injected into the soil at selected points, or in lines from a moving ploughshare-like device, or they can be applied as a surface treatment allowing natural precipitation to bring them down to the root level. Supplementary irrigation may be required to leach the compound down to the roots. As compounds tend to be broken down rapidly by the soil micro-organisms and chelated by the soil colloids, comparatively large doses are required, and even then are ineffective against a large number of diseases.

Aerial parts of a plant are more accessible and can therefore be treated more easily and cheaply. Most fungicides and bacteriosides do not travel very far from the point of application in the plant so that only diseases affecting the shoots can be treated by aerial application of these compounds. Recently, however, there has been the development of systemic fungicides that are transported away from the site of application throughout the plant. So for the first time root infections can be treated without the trouble and expense of soil application.

The dose of a compound applied to a plant must be such that it is effective against the pathogen without being toxic to the host. This can be achieved in a number of ways. Aerial parts can be sprayed with a dilute solution of the fungicide or bacteriocide, plus a wetting agent to the point of run-off. This means the treatment is sprayed on in large quantities and the excess runs off. Theoretically, the amount retained should be proportional to the leaf area, so that a controlled dose results but in practice it is difficult to obtain uniform coverage, particularly to the underside of leaves. The equipment required for this method of application is simple, ranging from a hand-pumped spray to a mechanically-driven, tractor-drawn spray. The disadvantage is that large quantities of water have to be transported and much of the treatment is wasted in the run-off. Dusting is also used as a method of application and can give more uniform coverage to the upper and lower surfaces, but adhesion to the plant and drifting on to neighbouring crops can be a problem. A more recent method of application is to use air as the diluent rather than water. A fine mist of concentrated solution is applied in an air-blast from a low-pressure compressor. The advantages of this method of application are that the blast of air moves the foliage about giving more uniform coverage and only small amounts of water need to be transported, making the equipment more portable. The disadvantage is that the dosage applied is dependent on the speed of movement of the applicator. This is not too difficult to control if it is a tractor-drawn device but a reasonable amount of care must be taken in the use of hand-held applicators to maintain a constant speed of movement. Other disadvantages are that the solution used is more concentrated, and therefore potentially more toxic, and there is a greater chance of inhalation of the preparation than occurs in spraying with large quantities of water. Drifting too can be more of a problem with air-blast spraying.

All these techniques have been adopted for use by aircraft, both fixed wing and helicopter, where large areas require treatment. The downdraft from the aircraft, particularly from helicopters, helps to promote uniform coverage by moving the vegetation about, but drifting can be even more of a problem with this method of application than with any of the others.

With all these methods climatic conditions must be considered. With root application heavy rain may leach the treatment down below root level before it is effective, or inadequate rain may leave it above root level. Rain is also important with foliar treatments as it may wash the compound off the leaves, but light rain can be beneficial in redistributing the fungicide or bacteriocide to areas not originally covered, thus increasing its effectiveness. Excessively dry air at the time of spraying can be a problem, especially with air-blast application, because the droplets may evaporate completely before reaching the plant so that the chemical arrives at the plant as a dry powder which may not adhere. Wind is also a problem with air-blast and dusting as it can reduce adhesion to the target plant and cause drifting to areas that should not be treated.

It is not always easy to understand the way in which a treatment acts. The application of a fungicide or bacteriocide to foliage prior to infection can give protection which may be due to inhibiting germination or to killing the pathogen after germination but before penetration or even killing the pathogen after penetration. Similarly it is difficult to tell if a post-infection treatment has killed the pathogen or merely halted its growth and whether this effect is due directly to the compound applied or to an alteration of the host metabolism, making it more resistant to infection. Although these questions are steadily being answered they are not of primary concern to the plant protection industry whose interest lies with the end result, disease control. The industry is mostly concerned with the pathogens against which the compound is effective, the conditions under which it is most effective, the best method and time of application, and any possible phytotoxic effects. For this reason large scale screening trials tend to be the first priority in the discovery and testing of new compounds, and it is only recently that our knowledge of the host-parasite interaction and of host and parasite physiology has been of use in the designing of new plant protection compounds.

Any fungicides or bacteriocide for general use should ideally meet all the following requirements: (i) It should be effective at a concentration which does no damage at all to the plant to which it is applied and should have a margin of safety so that errors in dilution and application have a minimal effect. (ii) It should not be poisonous to man or animals. (iii) It should be chemically stable and physically tenacious enough to be active for a considerable period of time after application. (iv) It should not adversely affect the ecology and micro-ecology of the area to which it is applied. (v) It should be of a type to which pathogens do not become resistant. (vi) It should be easy to apply in an accurate and effective manner, i.e. it should either be easily dissolved or should be as a finely-divided powder that does not form lumps when suspended in water (or compact into lumps in the case of dusting powders). (vii) It should be compatible with other pesticides so that a mixed treatment can be given. (viii) It must have a long storage life. (ix) It should be cheap in relation to the value of the crop.

As yet no fungicide or bacteriocide meets all these exacting requirements but enormous improvements have been made in recent years. One of the most exciting directions is in the field of systemic fungicides which as yet are still in their infancy but their development could well revolutionize plant protection.

7.3 Main groups of plant protection compounds

A very large percentage of the commercially important plant diseases are caused by fungi and therefore it is against these that most of the effort has been directed and most of the success achieved in chemical crop protection.

Copper

One of the first effective fungicides, applied originally as Bordeaux Mixture (approximately 5 parts copper sulphate to 7·5 parts slaked lime in 450 parts water). This results in a whole range of complex copper and calcium carbonates, hydroxides and sulphates which vary in type and proportion depending on the mixing procedure and conditions. The chemistry of this mixture as well as its mode of action has still not been fully elucidated. A variety of other copper compounds are now used as well as Bordeaux Mixture. All the copper-based fungicides have a wide spectrum, being active against ascomycetes, fungi imperfecti, downy mildews and bacteria.

Mercury

Many mercurial compounds are effective as fungicides and bacteriocides but comparatively few are used because of their mammalian toxicity. They are used mainly as seed dressings and wood preservatives. Inorganic mercurials have now been superseded by organo-mercurials, most of which have the formula $R-Hg-X$ where R is an organic radical (alkyl, aryl, alkozo-alkyl etc) and X is an acid radical, either organic or inorganic.

Other organo-metallics

Organic compounds based on tin, lead, germanium, arsenic, antimony, bismuth and several other metals have all been found to be effective under certain conditions. Those based on tin seem to be the most effective and still have tolerably low human toxicity but they tend to show fairly high phytotoxicity to some plants.

Sulphur

Finely divided sulphur was probably the first fungicide used, and is also useful in controlling some mites. Lime sulphur has similar properties and shows better adhesion to the plant. It is safe to use but is effective only against a limited number of diseases, including smuts and powdery mildews. Organo-sulphur compounds are more effective than the inorganic forms and are less likely than lime-sulphur to cause leaf scorch. For example Thiram (tetramethylthiuram disulphide) is used as a

Figure 7.1 Thiram

seed dressing against 'damping off' and as a foliar spray (except on fruit which it will taint). It is of low toxicity but can cause skin irritation. Interestingly it was first produced as a rubber curing accelerator before its fungicidal properties were discovered. The soil fumigant Dazomet is a thiadiazine which is incorporated into the soil as a dry powder where it breaks down liberating a fumigant gas.

Metalo-sulphurs

These attempt to combine the effectiveness of organo-metallic compounds and organo-sulphur compounds in the form of dithiocarbamates. Zinc and iron are the most effective metals in such compounds and are produced commercially as *Ziram* and *Ferban* which are effective against 'damping off', ascomycete and fungi imperfecti as well as some rusts and downy mildews.

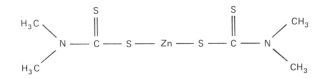

Figure 7.2 Ziram

Variations on this formula include *Zineb* in which the two ends of the molecule are folded round to form a ring structure and *Maneb* which is similar to Zineb except the zinc is replaced by manganese.

Non-metallic compounds

The most widely used of this type of fungicide is Captan which has a particularly broad spectrum of action but is not effective against rusts and mildews. There are several other useful fungicides in this group including *Glyodin, Oxine* and *Folpet* which will control a limited range of pathogens.

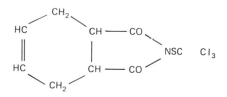

Figure 7.3 Captan

Although widely implicated in natural host resistance only two quinones have been found to be effective in field conditions. The two used are Chloranil and Dichlone The former is of some use on legumes but its application is limited because it decomposes in light, and the latter cannot be used on some plants because of phytotoxicity.

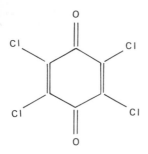

Figure 7.4 Chloranil

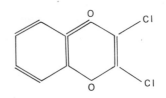

Figure 7.5 Dichlone

There are a number of aromatic fungicides, one of the most useful of which is *Dinocap*. Developed as an agaricide it is also effective against powdery mildews. Chloro-nitrobenzenes and chloro-nitroanilines are used for specific applications but their spectrum is very narrow.

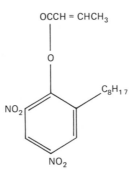

Figure 7.6 Dinocap

The aliphatic group of compounds, as well as being used as soil sterilants, also contain a number of members showing fungicidal properties, *Dodine* (*n*-dodecyl-guanidine) is effective against ascomycetes, such as apple scab, and many of the fungi imperfecti, but its usefulness is limited by its toxicity to some plants.

Antibiotics

Most fungal pathogens are insensitive to the antibiotics we have at present but the latter are used to control some bacterial plant diseases. The most useful ones in this respect are streptomycin and oxytetracyline which can be used to control fire blight of apple and pear, halo blight of bean and crown gall of a number of crops.

Systemic Compounds

We have briefly considered the protective role of fungicides but not their therapeutic action, i.e. the destruction of the pathogen after it has established itself in the host. The compounds previously mentioned may have a therapeutic effect as well as a protective one but this will only occur where the fungicide is able to reach the site of infection. As these fungicides are translocated very little, therapeutic effects can only occur in surface infections. For a compound to be really efficient in its control of an established infection it must be translocated throughout plant.

Because systemic compounds are transported to all parts of the treated plant it is not essential for their application to be made at the site of infection but it is generally advantageous to do so as near as possible so that the maximum concentration of fungicide is at or near the site of heaviest infection. Ease and efficiency of application may outweigh this consideration so that foliar sprays are sometimes used to protect against root infections and seed dressings which are absorbed by the roots may be used to protect the aerial parts of a seedling. It is for this reason that systemic fungicides are usually applied in one of these two ways, but there are disadvantages even with these methods of application. Foliar sprays may dry before absorption, or may be washed off by rain, and seed dressings may be destroyed by microbial or chemical action, or might become chelated by soil colloids before they are taken up by the plant. These disadvantages apply to any plant protectant and the greater efficacy of systemic compounds at treating the entire plant, especially in regions to which direct application cannot be made, is one of the most exciting developments of plant protection in recent years.

The best known and most widely used systemic fungicide is *Benomyl* which is effective against a wide range of ascomycetes (including powdery mildews), some fungi imperfecti and has even given good control of some basidiomycetes. It is, however, of little use against phycomycetes.

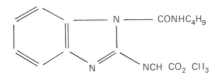

Figure 7.7 Benomyl

Other benzimidazole compounds are being investigated and some have been found to be useful as systemic fungicides, including one originally developed as a helminthocide.

The oxathiins form another group of compounds with members having systemic fungicidal properties. These are particularly effective against rusts and smuts, one compound being regularly used to free barley seed from loose smut present inside the seed.

A highly specific systemic fungicide which is very useful against some powdery mildews but is ineffective against others is *Dimethirimol*. This and its related compound *Ethirimol,* control powdery mildew on cucurbits and cereals but are ineffective against powdery mildew of roses.

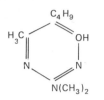

Figure 7.8 Dimethirimol   Figure 7.9 Ethirimol

There is a growing group of compounds which are systemic and control the pathogen within the host but which are not fungicidal in tests *in vitro*. These may act by being converted into fungicidal compounds within the plant or by altering the metabolism of the plant making it more resistant.

The dearth of knowledge about how fungicides and bacteriocides act within a plant and how this combines with a plant's natural resistance mechanisms makes the design and discovery of new and better plant protectants extremely difficult and there is a considerable element of trial and error in such investigations. We must therefore work for a much better understanding of the host-parasite relationship so that plant protectants can be accurately designed for a specific mode of action.

7.4   Epidemiological effects of fungicides

Fungicides are epidemiologically no different from any of the other factors we have considered. Where foliar sprays inhibit the germination and penetration of spores they can be considered in exactly the same way as any other environmental condition. It makes no difference to the infection rate if spores are killed by a fungicide, desiccation, ultra-violet in sunlight or host exudates. In this context fungicides could almost be considered a form of sanitation, with the exception that they may be applied at any time in the development of the disease, not just at the beginning. The action of fungicides, particularly systemic fungicides, in slowing down the growth of the pathogen is epidemiologically somewhat different to its sanitary action. In this role the fungicide effectively increases the resistance of the host and reduces the rate of development (r) of the disease.

In considering the effectiveness and action of fungicides we should consider as well as the pathogen and host, the environmental conditions. The weather, mineral nutrition and any other diseases, such as viral or bacterial infection from which the host may be suffering will all alter the effectiveness of the fungicide. For this reason results, particularly those of field trials, tend to vary and a fungicide which may be highly effective against a particular pathogen under one set of conditions may be totally ineffective under others. Similarly a treatment may cause no damage to the host in one situation and be highly phytotoxic in others. Plant protectants have therefore to be tested over a wide range of conditions, particular attention being paid to their effectiveness under conditions which would normally cause a major epidemic of the disease against which it is being tested.

Let us consider the results of a particular experiment using a copper fungicide. Tomato plants grown in a greenhouse were sprayed with copper salts under as near ideal conditions of application as it was possible to obtain. The plants were then inoculated with a suspension of *Phytophthora infestans* spores and after a period of incubation the number of lesions developing was counted. Under these conditions 99 per cent of the spores were killed before they could establish an infection. We can therefore take 99 per cent as about the highest kill that copper is likely to achieve against *P. infestans*. Other results confirm these findings. Detached potato leaves were similarly treated prior to inoculation and a 90 per cent kill was achieved. What effect would these treatments have on an epidemic?

$$\Delta t = \frac{2 \cdot 3}{\text{infection rate}} \times \log_{10} \text{sanitation ratio}$$

where; Δt = time taken to reach the same level of infection as an untreated crop

In these instances with an infection rate of 0·23

$$\text{For 90 per cent kill} \quad \Delta t = \frac{2 \cdot 8}{0 \cdot 23} \times \log_{10} \frac{100}{10}$$

$$= 10 \text{ days}$$

$$\text{For 99 per cent kill} \quad \Delta t = \frac{2 \cdot 3}{0 \cdot 23} \times \log_{10} \frac{100}{1}$$

$$= 20 \text{ days}$$

This means the copper treatment would set back the course of the disease ten or twenty days.

This assumes that the fungicide is effective only against the first infection and does not alter the growth rate of the pathogen following infection. Happily the fungicide is often effective for longer than this initial period and may well slow down the growth of the fungus within the host for some time, but repeated spraying is still often necessary. In one experiment unsprayed potato crops had a very high infection rate with a rapid build-up of *P. infestans*. Spraying with fungicide once or twice during the season markedly reduced this rate of build-up and spraying between three and five times drastically reduced the infection rate of the pathogen. Clearly repeated application will have more impact on the infection rate than a single application but even a single application will reduce the infection rate for a limited period of time. Different types of fungicide will vary in the time for which they are effective and fungicides which are active for a long period of time may well be more successful in controlling a pathogen than a short-acting compound even though the long-acting compound may give a lower initial kill of the pathogen.

In crops with a low infection rate disease control using fungicides is easier than in those with a high rate. In some cases the rate is low enough for disease control to be virtually complete. For example in citrus black spot caused by the ascomycete *Guignardia* the young fruits are infected by ascospores released from fallen infected leaves, but it takes several months from the time of infection for sufficient damage to have been caused to the fruit for the lesion to be visible to the naked eye. Spraying with copper fungicides is carried out at the time of petal fall from the blossom and then two further treatments at six-week intervals. At the time of the last spraying the fruits are still tiny so that during the following weeks the fruit swells, increasing the surface area to many times its size at the time of spraying. Despite this further dilution of the fungicide there is usually a reduction of about 98 per cent in the number of infection sites, which may be due to an extreme sensitivity to copper on the part of the pathogen or to the long period of time the pathogen spends on or near the surface of the fruit exposed to the copper fungicide.

Similarly apple scab caused by *Venturia inaequalis* can be controlled right to the end of the growing season by the application of fungicides. Two of the most widely used Captan and Dodine, form a protective barrier against infection and they reduce the number of spores produced by the pathogen on tissue that does become infected. Furthermore those spores that are produced have reduced infectivity. The first action reduces the original inoculum x_0, and is therefore a sanitary effect, whilst the latter actions reduce the rate of spread of the disease r.

The following table shows the calculated percentage reduction in viable spores required to stop an epidemic, that is, to reduce iR_c to less than one (where i = time during which infections liberate spores and R_c = number of spores liberated per unit time):

Table 7.1 *Percentage reduction in spores (kill) required to halt epidemic.*

r (per unit per day)	i = 1 day	i = 0·75 day
0·5	89	84
0·4	83	76
0·3	74	65
0·2	59	46
0·1	36	15

(the latent period, p, taken as 4 days)

This clearly demonstrates that the greater the rate of spread of the disease and the longer the period over which spores are shed the more difficult it is to contain it by fungicidal treatment and that fungicides which by their action reduce r have a far greater likelihood of being successful in controlling an epidemic.

The time of application can also be important in that is is generally easier to avoid an epidemic than to stop it once it has started. It is therefore generally agreed that spraying with a fungicide should take place just before the first infection is likely so that it has the maximum sanitation effect. It is difficult to predict when infection is likely to occur and as spraying is expensive both in the cost of the compound applied and in the labour of application it is undesirable to treat a crop that is not at risk. With highly susceptible crops in which there is a very rapid increase in the pathogen there is little choice but to spray early and to continue regular treatment throughout the season. With other crops only at risk in certain climatic conditions meteorology can be used to warn farmers of times when crop protection will be required and 'blight warnings' for instance may be issued. In crops with a comparatively slow spread of the pathogen treatment at the first symptoms of infection may be quite adequate. For example *Phytophthora* blight may be controlled on some varieties of potato by a single spraying of copper oxychloride at the first sign of the disease.

7.5 Breeding for disease resistance

To be resistant to infection is obviously of advantage to an individual plant but how is it of evolutionary advantage to the species? A pathogen will only affect the evolution of a species if it alters the ability of the members of the species to reproduce and propagate. A pathogen which does not affect a plant's reproductive ability will therefore exert no evolutionary pressure. The greater the damage to the host species the greater the selection pressure to evolve resistant strains. Clearly in a species being decimated by a pathogen a resistant individual will be at an enormous competitive advantage and will oust the susceptible varieties from their ecological niche, further increasing the pressure on the susceptible varieties which may well result in them becoming extinct. It is for this reason that wild plant species show a high degree of resistance to a wide variety of pathogens, and in general the successful pathogens are highly specialized, causing comparatively little damage to the host or causing major damage only after reproduction has been accomplished.

In crop varieties the situation is rather different. For hundreds of generations plants have been selected for high yield to the exclusion of practically everything else, with the result there has in some cases been a positive selection of disease-susceptible plants and almost invariably the natural selection in favour of resistant specimens has been halted. You may ask why halting the continued natural selection of disease-resistant specimens should lead to an actual increase in susceptibility. This is because the balance between host and parasite is a dynamic one both at the physiological and evolutionary level. The host plant constitutes a major part of the environment in which the pathogen lives. If an individual pathogenic organism is more successful in growing in its host and produces more propagules than its neighbours, its descendants will become the dominant race of that species of pathogen. In a natural situation this will of course increase the selection pressure on the host to evolve strains resistant to this more successful race of pathogen. A dynamic balance will thus be maintained. Agriculture has removed the ability of the crop to respond to these changing pressures.

There is also another difference between crop plants and wild plants. Damage done to wild plants that does not affect their survival and reproductive abilities can be neglected but similar damage done to crop plants may make them unmarketable. So that some pathogens may cause damage that is biologically unimportant but which is commercially disastrous.

To combat these two effects which can make the disease situation in crops so serious, there has been an increasing emphasis on plant breeding, not only to produce plants with a high yield, but also to produce them with high levels of resistance to the pathogens particularly serious in that crop. There are a number of approaches to this problem. The two most common are the screening of commercial varieties for disease-resistant strains, or in some cases even resistant individuals from which to breed a disease-resistant variety, and the crossing of commercial high-yielding varieties with disease-resistant wild varieties followed by a screening programme to select those individuals with a combination of desirable characteristics. Both these methods have difficulties. The first assumes that in the population of commercial plants there are some members retaining a fairly high degree of disease resistance. This may be true in species that have not been intensively bred and therefore have wide range of genes in the population, it is unlikely to be true in intensively selected species which would have a narrow genetic base. The second method has difficulties because the act of broadening the genetic base of a species by incorporating genes from a wild variety frequently reduces the yield, the reverse of the ultimate objective of the agriculturalist. All breeding programmes run into the problem that there are frequently many resistance mechanisms operating in a

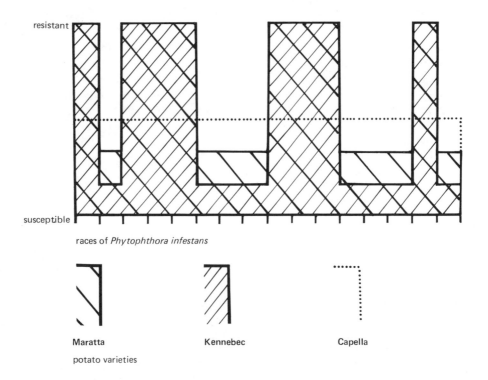

Figure 7.10 Comparing the high levels of resistance to specific races of *Phytophthora infestans*, and the low level of resistance to other races, shown by 'Kennebec' potatoes, with the high-levels of resistance to some races and moderate levels of resistance to other races shown by 'Maratta' potatoes, and the fairly high level of resistance shown to all races of *Phytophthora infestans* by 'Capella' potatoes. (redrawn from Van Der Plank)

single individual, one of which may be important in one environmental condition whilst another may be important in a different environmental condition. This means that selection for disease resistance in one environment may result in selection for susceptibility under different conditions. A similar effect can also take place in that breeding for resistance to one disease may select for susceptibility to another disease.

These pessimistic statements do not mean that breeding for disease resistance is impossible; quite the reverse: it is possible and the potential gain is enormous. They do mean that breeding for disease resistance is difficult, time-consuming, expensive and can only be done by well-designed experiments and trials. Considerable progress has already been made in the breeding of a number of crops with resistance to several diseases. The results so far seem to indicate that genetically there are two sorts of resistance: (i) practically complete resistance to certain races of a pathogen and a high degree of susceptibility to other races; (ii) moderate levels of resistance to all the races of a particular species of pathogen. The first sort is called *vertical resistance* and appears to be fairly simply inherited, in some instances the inheritance being monogenetic. The second form is called *horizontal resistance* and is usually inherited in a complex manner involving a large number of genes.

Breeding for vertical resistance is very much easier than breeding for horizontal resistance. This is because the monogenetic or aligogenetic system involved in vertical resistance is far easier to handle than the polygenetic system of horizontal resistance. Vertical resistance, in addition, is far more obvious than horizontal resistance, a vertically resistant plant showing almost complete resistance to the pathogen whereas the horizontally resistant plant may only show a slight increase in resistance which can easily be missed due to variations in environmental conditions.

The easier breeding programme and the more dramatic results produced by vertical resistance have led plant breeders to use this as the basis for their development of resistant varieties. The disadvantage is that just as in breeding for high yield it is easy to select disease susceptible plants, so in breeding for vertical resistance it is easy to select accidentally plants with very low levels of horizontal resistance. This means that a crop with a high level of resistance to one race of a pathogen will have little resistance to others. Thus, in an area where one particular race of a pathogen is endemic vertical resistance can provide almost complete protection against the disease, but should another race become introduced from outside the area, the crops will be exceedingly susceptible and a disease epidemic is likely. Even if a crop is selected for vertical resistance to all known races of a pathogen it will still be liable to attack by new races produced by mutation or genetic reassortment. In fact, the breeding of vertically resistant crops provides an ideal environment to select a new virulent race of the pathogen because if one does arise it will meet with little host resistance so will produce large numbers of propagules which will have little or no competition from the races unable to . colonize the host. In a comparatively short period the new race will be the major representative of that species of pathogen in that area. This means that plant breeders must be continually breeding new varieties resistant to these new races, an activity which in itself is helping to select still newer races of the pathogen.

The time scale of these events depends on the genetic plasticity of the pathogen. With some diseases the production of new races of the pathogen may be so infrequent that it can be neglected, in which case vertical resistance is an effective and invaluable form of disease control. With other diseases the pathogen in continually producing new races, most of which will be less successful than the established races but a tiny percentage will be of greater vigour on the resistant host types and will rapidly spread through the resistant crops. With such diseases the plant breeder is always half a step behind the pathogen because he does not know which race of the pathogen against which to breed for resistance until they have appeared and are spreading rapidly. In such cases vertical resistance is of far less use.

Horizontal resistance is much less dramatic than vertical resistance both in its successes and failures. Horizontal resistance will very rarely provide complete protection against a pathogen but it does limit its rate of spread and reduces the damage done by it. More importantly, it is equally effective against all races of a particular pathogen so that there is not the risk of a new race, either from outside the area or produced by mutation and genetic reassortment within the area, causing an epidemic. Nor does it produce a selection pressure favouring the spread of new races of a pathogen.

Obviously the ideal solution is to have a crop with vertical resistance to all the known races of a pathogen coupled with a high level of horizontal resistance. As previously mentioned, breeding for horizontal resistance is more difficult than for vertical resistance and the combination of the two makes breeding still more difficult. However the potential gain, particularly against genetically very plastic pathogens is enormous, so that breeders have for some years now been working towards this goal and in a few instances achieved some remarkable successes.

The physiological differences between vertical and horizontal resistance are not yet completely clear. Vertical resistance is easily understood in the case of highly obligate parasites which require some part of the host's metabolic pathways or complex metabolites for its survival. If these pathways or metabolites are even slightly different the pathogens may not be able to utilize them and the plant will be resistant, that is, until a race of pathogen comes along which is able to make use of the new metabolite or pathway. To such a pathogen the plant will be highly susceptible. Horizontal resistance is also easy to understand where mechanical barriers are involved, and all races of a pathogen will meet with a similar obstacle.

The situation unfortunately is not always so simple. Many plants show vertical resistance to pathogens able to grow on very simple media, so nutritional requirements are not the only cause of vertical resistance. Vertical resistance seems to be related to the hypersensitive reaction, resistant plants undergoing a more violent reaction and producing larger amounts of phytoalexin, causing the rapid death of any pathogen triggering this reaction. Horizontal resistance appears to be involved not only in the production of mechanical barriers before and after infection but also with the production of anti-microbial chemicals before and after infection. The production of mechanical barriers ahead of infection and the anti-microbial chemical deposited in them is a more complex process involving more intercellular co-operation than the hypersensitive reaction occurring within one or two cells at the site of infection. This is perhaps reflected in the genetic complexity with which horizontal resistance mechanisms are inherited compared to the relative simplicity of the inheritance of vertical resistance.

7.6   The epidemiology of resistance

The epidemiological effects of resistance are to reduce the original effective inoculum and to reduce the rate of development of the disease. Vertical resistance reduces the effective inoculum of a pathogen because usually there are propagules of a large number of races of a pathogen challenging the host's defence mechanisms. If the plant has vertical resistance to many of these races it will be practically totally resistant to them and epidemiologically we can neglect them. We then need only consider those races of a pathogen to which the plant shows at least some degree of susceptibility. If a plant is susceptible to only one spore in ten falling upon it, this is epidemiologically the same as reducing the number of incident spores to one tenth, i.e. the original inoculum x_O has been reduced by 90 per cent. The problem is that if large amounts of this variety of crop are grown, the majority of the spores or propagules will ultimately be of the virulent type. The time taken for this to happen may be long or it may be short, but eventually a new equilibrium will be set up between the various races of the pathogen and the position of this equilibrium depends on the amount of the resistant host variety grown in that area. If a tiny amount is grown the multiplication of the virulent race will necessarily be small and the virulent propagules will be diluted with a vast number of propagules non-infectious to the resistant plant, so the latter will remain comparatively disease free. If the percentage of resistant crops grown in one area increases then so will the percentage of the pathogen virulent to them, so that the maximum benefit will be obtained by a farmer growing a crop showing vertical resistance whilst his neighbours grow susceptible varieties. As soon as they start to grow the same resistant variety then the amount of benefit to the original farmer will start to decline.

   This is clearly unsatisfactory and to overcome this problem crops showing vertical resistance to all the races of a pathogen in that area can be grown by all the farmers in that area with benefit to them all. That is until a race of the pathogen to which the crops are not resistant occurs in that area. This can be due to movement of the pathogen from one part of the world to another, to spontaneous mutation of the pathogen or to re-assortment of the genetic resources of the pathogen population. The first of these causes of resistance failure is reduced by restrictions on the movement of plant material and soil debris about the world. If trade is to continue this cannot be totally successful and in any case it will not halt the movement of spores by wind and water. The second reason for failure varies with the genetic stability of the pathogen, and happily does not seem to happen very often in most species of pathogen. It is, however, alarmingly high in a few species and little or nothing can be done to stop it. Genetic re-assortment is much more important and for it to occur the sexual stage of the pathogen's life cycle is usually required. If the sexual stage of the pathogen is on some host other than the crop, the re-assortment can be reduced by removing the alternative host. For example, *Puccinia graminis* undergoes the sexual stage of its life cycle on barberry. So removal of barberry bushes will reduce the genetic re-assortment in *Puccinia* and decreases the likelihood of new races of the pathogen arising to cause black stem rust on wheat varieties showing vertical resistance to the previous races of the pathogen.

Vertical resistance effectively reduces x_O but horizontal resistance reduces the rate of growth of the pathogen r. As noted in the previous discussion of epidemiology, reductions in x_O are at their most effective where r is low, so that sanitation and vertical resistance will be at their most effective in the presence of high levels of horizontal resistance. Horizontal resistance alone will reduce the level of infection by a greater or lesser amount depending on the degree of resistance but it will be fairly constant from year to year. Obviously some years will be worse than others, depending on the weather conditions amongst other things, but these variations will be small compared to the fluctuations that would occur in a crop with considerable vertical resistance and little horizontal resistance. In such a crop there is likely to be fluctuation between no disease some years and epidemics other years because of the build-up of a virulent race of pathogen. An example of this is the black stem rust of wheat in N. America. There have been five major epidemics since 1885, when records were first started. These were in 1904, 1916, 1935, 1953 and 1954. The commercial use of vertically resistant wheat varieties become significant in about 1930. In the forty-five years prior to this date there were two major epidemics and in the forty-five years since then three major epidemics. Despite this seeming contradiction, more epidemics since the introduction of resistant varieties, there has been a sufficient reduction in the level of infection in the non-epidemic years to make it worth while. If the horizontal resistance of the new varieties had been higher the epidemics could perhaps have been averted.

It is therefore clear that the present aim of plant breeders is to produce high-yielding varieties, vertically resistant to a large number of races of a large number of pathogens coupled with a high level of horizontal resistance to these various pathogens.

7.7 The inter-relationship of disease control measures

New fungicides and bacteriocides that are cheaper and better than those available at present will no doubt be discovered in the future, but their cost and the labour involved in their application will still be significant, particularly in the lower value crops. We must not therefore lose sight of the other methods of control. Sanitation can in suitable circumstances retard the development of a disease, in some cases sufficiently to make other control measures unnecessary. Even when it is not adequate it will reduce the period over which treatment with chemical protectants is needed as well as making the application more effective. Sanitation also has the advantage of no possible phytotoxicity or animal toxicity.

Breeding for resistance has already done much for agriculture and is likely to do a great deal more in the future. Horizontal resistance can be coupled with other forms of control far more effectively than can vertical resistance. This is because if a vertically resistant crop is attacked by a race of pathogen to which it is totally resistant then fungicides or bacteriocides are unnecessary, and if it is attacked by a race to which it is highly susceptible then it is unlikely that a chemical protectant could be applied in time to save it from the explosive growth of the pathogen that would ensue. Horizontal resistance on the other hand can be easily and effectively coupled with other methods of disease control. Chemical protectants need to be applied at the right time, preferably only when required, and horizontal resistance can slow the course of infection sufficiently to make it practicable to control a disease even after the first symptoms are visible. We have previously noted that sanitation is at its most effective when the growth rate of the pathogen is low, so that a combination of horizontal resistance which slows the rate of growth of the pathogen, and sanitation which reduces the original infection to a low level, for a highly effective means of disease control. Any method by which the growth rate of the pathogen can be kept at a low level will enable sanitation to be effective, which means that the many chemical disease-control compounds that reduce the growth rate of the pathogen can be successfully coupled with sanitation.

Plant breeding has in the past tended to produce plants resistant to one particular disease, be it fungal, bacterial or viral. There is now evidence that some resistance mechanisms may be effective against a number of diseases and the production of crop varieties with a resistance mechanism giving protection against a wide range of diseases may not be too far in the future. Plant breeding may in the future make other plant protection methods obsolete but this is a long way off and it is probable that many pathogens will continually produce new races. This will keep pace with the plant breeders best endeavours so that probably all the methods of disease control will continue to be required to obtain the maximum protection with the minimum cost and the minimum undesirable side effects.

Glossary

Apothecium A disc-like structure on which asci containing ascospores are borne.

Ascospores Spores produced inside a sac (ascus). The sexual stage in the life cycle of ascomycetes.

Basidiospores Spores produced on the end of basidia in the sexual stage of basidiomycetes.

Callose A carbohydrate produced by plant cells to block pores in cell walls e.g. sieve plates and to inhibit invasion by some pathogens.

Callus Superficial unspecialized tissue produced by plants in response to wounding.

Cambium A layer of actively dividing cells producing new tissues e.g. fasicular producing new vascular tissue, cork cambian producing cork tissue.

Chlamydospores Spores with thick cell walls developing from vegetative cells in the middle of hyphal strands.

Cleistothecium A totally closed structure containing ascospores with their asci. Produced as the sexual stage of some ascomycetes.

Conidia Asexual spores borne at the tips of hyphae not in a sporangium.

Epinasty More rapid growth on the upper side of a leaf or petiole causing downward curving and twisting of the organ.

Epiphyte A plant attached to another plant but not growing parasitically on it.

Lignified Cell walls strengthened and thickened with aromatic compounds. Found in woody tissues.

Lignituber Deposits of lignin, cellulose and pectin around a hypha attempting to penetrate a cell. Part of a defence mechanism by host.

Micro-sclerotia A small knot of thick-walled vegetative hyphal cells produced as a resting structure by some fungi.

Middle lamella A layer of pectic substances on which the other cell wall materials are deposited. Forms the 'cement' between plant cells.

Mycorrhiza An association between a fungus and the roots of a host plant, often to the benefit of the host.

Perithecium A flask-shaped structure containing the asci and ascospores produced by some ascomycetes during the sexual stage of their life cycle.

Propagule Any structure specialized for the reproduction of the species. Usually small.

Phytoalexin Any antimicrobial compound produced by a plant in response to attack that is not present in a healthy plant.

Resistance An ability on the part of the host to slow down or stop the attack of a pathogen.

Rhizomorph A compact strand of fungal hyphae capable of transporting nutrients and spreading infection over considerable distances.

Rhizosphere Zone immediately around plant roots in which the microbial population is enhanced by root exudates.

Saprophyte A plant using dead organic material as its energy source.

Sclerotia Knots of thick-walled vegetative hyphae produced as the resting stage of some fungi.

Sieve tube Plant cells specialized for the transport of sugars, connected one to another by perforated sieve plates. Found in the vascular strands of leaves, stems and roots.

Suberin A mixture of fatty acid derivatives deposited in cell walls to make them impermeable to water, as in corky cells. Often produced in wound healing reaction and to prevent penetration by a pathogen.

Symbiosis Two or more dissimilar organisms living together to their mutual advantage.

Synergism A situation where the combined effect of two or more compounds or organisms is more than the sum of their individual effects.

Tracheid A dead lignified cell in the vascular strands of plants which conducts water and provides mechanical strength.

Tylose A bladder-like outgrowth from a living cell into the lumen of a wood vessel often causing complete blockage of the vessel.

Wood vessel A series of dead, lignified cells with the end walls broken down to form a continuous tube for the conduction of water in the plant's vascular system.

Zoospores A motile spore with one or two flagella.

Zygospore A thick-walled resting spore produced as the result of sexual reproduction by some phycomycetes.

Acknowledgements

I would like to acknowledge the following as authors of figures and tables used in this book.

BRAUN A.C. 1955, *Phytopathology*, **45**, 659–64

BULLER A.H.R. 1922, *Researches on fungi*, Vol.2. London

CARTER M.V. 1965, *Aust. J. Agric. Res.*, **16**, 825–36

DALY J.M. & INMAN R.E. 1958, *Phytopathology*, **48**, 91–7

GREGORY P.H. 1961, *The microbiology of the atmosphere*. Hill, London

HIRST J.M. 1958, *Outl. Agric.* **2**, 16

HIRST J.M. & HIRST G.W. 1967, *J. Gen. Microbiol*, **48**, 357–77

INGOLD C.T. 1971, *Fungal spores, their liberation and dispersal*. Clarendon Press, Oxford

ITO S. & SHIMADA S. 1937, *Contrib. Improvement. Agr. Ministry Agr & Forestry*, **120**, 1–109

KAWAMURA E. & ONO K. 1948, *Bull. Natl. Agr. Expt. Sta.* (Japan), **4**, 13–22

KRUPA L. 1958, *Science*, **128**, 477–78

LEE S. & LeTOURNEAU D.J. 1958, *Phytopathology*, **48**, 268–74

MAINE E.C. 1960, Thesis, Dept. Plant Path. North Carolina State Col. Quoted by, Goodman, Kiraly, & Zaitlin in *The biochemistry and physiology of infectious plant disease*. Van Nostrand 1967

McLEAN J.G., LeTOURNEAU D.J. & GUTHRIE J.W. 1961, *Phytopathology*, **51**, 84–9

MELANDER L.W. & CRAIGIE J.H. 1927, *Phytopathology*, **17**, 95–114

PILET P.E. 1960, *Phytopathol. Z.* **56**, 76–90

SHAW M. & SAMBORSKI D.J. 1957 *Can. J. Botany*, **35**, 380–407

STEPHANOV K.M. 1935, *Bull. Pl. Prot. Lenigr.* ser 2, *Phytopathology*, **8**, 1

SUZUKI N. 1957 *Bull. Nat. Inst. Agric. Sci.* (Japan) Ser. C. **8**, 69–130

VAN ARSDEL 1965, *Phytopathology*, **55**, 945–50

VAN DER PLANK J.E. 1963, *Plant disease; epidemics and control*. Academic Press

YOSHII H. 1936 *Ann. Phytopath. Soc.* (Japan), **6**, 199. 1937, Abstract; *Rev. appl. Mycol* **16**, 339

ZAITLIN M. & JAGENDORF 1960, *Virology*, **12**, 477–86

Index

Italicized page entries refer to diagrams